Prospecting Field Tests For The Common Metals

by Arizona Bureau of Mines

with an introduction by Kerby Jackson

Introduction

It has been decades since the Arizona Bureau of Mines released their important publication "Field Tests For Common Metals. First released in 1942, this important volume has been out of print and has been unavailable to the mining community since those days, with the exception of expensive original collector's copies and poorly produced digital editions.

It has often been said that *"gold is where you find it"*, but even beginning prospectors understand that their chances for finding something of value in the earth or in the streams of the Golden West are dramatically increased by going back to those places where gold and other minerals were once mined by our forerunners. Despite this, much of the contemporary information on local mining history that is currently available is mostly a result of mere local folklore and persistent rumors of major strikes, the details and facts of which, have long been distorted. Long gone are the old timers and with them, the days of first hand knowledge of the mines of the area and how they operated. Also long gone are most of their notes, their assay reports, their mine maps and personal scrapbooks, along with most of the surveys and reports that were performed for them by private and government geologists. Even published books such as this one are often retired to the local landfill or backyard burn pile by the descendents of those old timers and disappear at an alarming rate. Despite the fact that we live in the so-called "Information Age" where information is supposedly only the push of a button on a keyboard away, true insight into mining properties remains illusive and hard to come by, even to those of us who seek out this sort of information as if our lives depend upon it. Without this type of information readily available to the average independent miner, there is little hope that our metal mining industry will ever recover.

This important volume and others like it, are being presented in their entirety again, in the hope that the average prospector will no longer stumble through the overgrown hills and the tailing strewn creeks without being well informed enough to have a chance to succeed at his ventures.

Kerby Jackson
Josephine County, Oregon
May 2016

PREFACE

This pamphlet has been compiled as a text to be used in the Arizona Bureau of Mines' extension lectures on "Field Tests for Metals." Many of the people attending these programs, especially those most interested, have had very little, if any, technical training in mineralogy, chemistry, and allied sciences. For that reason this pamphlet has been written in as simple and non-technical language as the subject permits, yet the directions given are full enough so that anyone should be able to understand how to do the work.

The method of procedure suggested for beginners is first to practice each specific test for each metal on a mineral that is positively known to contain that metal or to practice the tests on a piece of that metal. This practice should be continued until the beginner becomes thoroughly familiar with the various tests for that metal as well as those for other metals. These tests can then be applied to ores of unknown metallic contents if it is desired to determine whether a certain metal is present.

TABLE OF CONTENTS

FIELD TESTS FOR THE COMMON METALS

By George R. Fansett

TESTING EQUIPMENT

Ultraviolet ray (black light) lamp
Blowpipe (8 or 10 inches long)
Heating apparatus (candle or alcohol, lard oil, or some other lamp)
Charcoal sticks (4 inches by 1 inch by ¾ inch)
Hand lens
Streak plate
2 or 3 inch magnet or magnetized knife blade
Jackknife
Forceps (platinum-tipped are the best)
3 pieces (No. 26 B. & S. wire gauge) platinum wire and holder for
    the wire
Iron wire (baling wire), a few pieces 4 inches long
Pyrex test tubes (4 inches by ½ inch or 6 inches by ¾ inch)
Porcelain (china) cup
Porcelain crucibles
Soft glass tubing, 7 mm. (for closed and open tubes). Some prefer
    to buy these tubes already manufactured rather than to
    make them.
Book of litmus paper
Window glass (few pieces for fluorine test)

DRY REAGENTS

¼ ounce powdered borax. Borax glass is more satisfactory than
    ordinary borax.
1 ounce powdered sodium carbonate (baking soda)
¼ ounce powdered salt of phosphorus (sodium metaphosphate)
4 ounces zinc metal (mossy or granulated. 20 mesh or finer)
2 ounces tin (shavings or filings)
1 ounce powdered manganese dioxide
¼ ounce sodium acid phosphate (for magnesium test)
¼ ounce diphenylamine crystals (for nitrate test)
1 ounce sodium bismuthate (for manganese test)
2 ounces sodium or potassium ethyl xanthate (for molybdenum
    test)
2 ounces potassium or sodium hydroxide (pellets) (for molyb-
    denum test)

WET REAGENTS

4 ounces concentrated hydrochloric (muriatic) acid
4 ounces concentrated nitric acid
2 ounces concentrated sulfuric acid (oil of vitriol)
8 ounces concentrated ammonia
3 ounces denatured alcohol

¼ ounce (10 per cent) cobalt nitrate solution
1 ounce dimethylglyoxime solution (for nickel test)

OTHER USEFUL TOOLS AND SUPPLIES

1½ inch agate mortar and pestle (some use a black iron pipe cap
    as a mortar)
Gold pan, horn spoon, or frying pan for panning
File (4 inch triangular)
Hammer
Anvil (block of steel 1½ inches by 1½ inches by ½ inch is con-
    venient)
Flame-color screen (Merwin's)
Small beakers (about 120 cc.)
2 2-inch glass funnels
1 package 4 inch filter papers
1 ounce potassium bisulfate (acid sulfate of potassium)
¼ ounce ammonium oxalate (for calcium test)
¼ ounce ammonium molybdate (for phosphate test)
1 ounce hydrogen peroxide
1 ounce of sulfuric ether or chloroform (for the petroleum test)
¼ ounce ferrous sulfate (for gold test)
¼ ounce stannous chloride (for gold test)
2 ounces mercury (quicksilver)
1 zinc stick
3 candles (for fluorine test)

BLOWPIPE OPERATIONS

G. M. Butler's *Handbook of Blowpipe Analysis* gives the following explanations of the flames made with the aid of the blowpipe and the directions for producing them:

The blowpipe is used for the purpose of concentrating the flame into a
long slender cone which can be readily directed against the substance to
be heated. It is very important that the blast be continuous and uniform,
although this operation may seem very difficult at first. The blast is not
produced by the lungs, but results from a bellows-like action of the dis-
tended cheeks. During the operation, air is inhaled only through the nose,
and is exhaled largely through the mouth and blowpipe. Before try-
ing to use that instrument, distend the cheeks, and keeping the mouth
closed, breathe through the nose for a moment; then open the lips just
enough to allow a little air to escape slowly, and admit air from the lungs
by a kind of gulping action just fast enough to keep the cheeks fully dis-
tended. This may take some practice, but, when it is possible to allow the
air to escape continuously from the mouth in this way, no matter whether
it is being exhaled or inhaled through the nostrils, it is time to begin to use
the blowpipe.

Producing the Oxidizing Flame. Place the oil lamp so that the longer
dimension of the wick is from right to left, and set its right-hand edge upon
a pencil or some other low support so that it will tip somewhat to the left.
Insert the tip of the blowpipe about one-eighth of an inch within and just
above the right-hand side of the wick, and blow steadily parallel to the
wick, directing the flame to the left, and producing a clear, blue flame
about an inch long. If all of the flame cannot be thus diverted to the left,
or if there are yellow streaks in the flame, trim or lower the wick. If the

whole flame is inclined to be yellow, move the tip of the blowpipe a trifle to the left. If it is impossible to produce a flame approaching the length mentioned above, the opening in the end of the blowpipe is too small, and this opening is too large when a very long, hissing flame is produced. In order to succeed in blowing a steady flame, the hand must rest upon some support, or the third and fourth fingers may be placed against the lamp.

In analytical operations, it is sometimes desirable to oxidze substances to be tested, and at other times the aim is to reduce them to the metallic condition; either result can be more or less readily obtained with the blowpipe.

A flame produced in the manner above described is called an oxidizing flame, but the action of all portions of such a flame is not oxidizing. The blue cone contains considerable carbon monoxide and is feebly reducing in its action, but just outside of the blue cone at the tip of the flame is an extremely hot, but nearly colorless, zone which is strongly oxidizing because of the free oxygen there present, and anything held in this zone about an eighth of an inch from the tip of the blue flame will be in the most favorable position for oxidation.

The oxidizing flame is hotter than the reducing, and the hottest part of this flame is just outside of the blue cone. In the absence of other instructions, substances should always be heated there.

Producing the Reducing Flame. Hold the tip of the blowpipe about one-sixteenth of an inch above and to the right of the wick, and a long, yellow flame containing much unconsumed carbon will be produced. This is sometimes called the smoky, reducing flame. Where greater heat is required, the inner cone of the oxidizing flame should be used. The strongest reducing action will take place at the tip of, and within the yellow cone of the reducing flame.

Note.—*One of the best kinds of lamps for this work is one burning a mixture of one third kerosine and two thirds lard oil, but the flame from a candle, an alcohol or any other kind of lamp, or from a Bunsen burner may be used.*

TESTS FOR METALS AND MINERALS

ANTIMONY

The most important antimony mineral of commerce is stibnite (antimonite, antimony glance, gray antimony, or antimony sulfide).

1. Antimony can usually be detected by the dense, white fumes given off and the heavy, white sublimate formed near the mineral when it is heated on charcoal before the blowpipe.

To make this test: Transfer to a flat stick of charcoal a little of the mineral to be tested for antimony. Use about the amount of powdered mineral that can be held on the tip of a knife blade or a piece of the mineral about as large as a kernel of wheat. Heat the mineral before the blowpipe in the oxidizing (bluish) flame until the material is thoroughly fused. By this treatment, most antimony minerals give off dense, white fumes which often continue to arise even after the heating has ceased. These fumes deposit as a dense, white sublimate[1] (coating) on the charcoal near the mineral, the outer edges, where the coating is thin, appearing bluish white.

Beginners practicing this test should use antimony sulfide.

2. Sulfides of antimony if heated in a closed tube yield a sublimate which is black when hot and reddish brown when cold.

To make this test: Fill a closed tube to about ½ inch from the bottom end with the powdered mineral to be tested. Heat the lower portion of the tube at a red heat for some time. Antimony sulfides if present in the mineral tested will give a sublimate (coating) on the walls of the tube. This coating is black when hot, but on cooling changes to reddish brown.

Beginners practicing this test should use antimony sulfide.

ARSENIC

Nearly all of the arsenic that is marketed in this country is obtained as a by-product from the fumes given off when smelting other ores.

1. When struck a glancing blow with a hammer, many arsenic minerals give off sparks and a garliclike odor.

Beginners practicing this test should use arsenopyrite.

2. Arsenides, sulfides of arsenic, and native arsenic give off a garliclike odor when heated on charcoal before the blowpipe. This treatment also gives a white coating on the charcoal that forms at a distance from the mineral.

---

[1]This bluish-white coating of antimony must not be confused with those of lead or zinc which are very similar on the outer edges. These coatings can easily be identified since the coating deposited on the charcoal from lead is yellow near the assay when hot or cold, and the coating deposited on the charcoal near the assay from zinc is yellow when hot and white when cold. The coating from zinc can also be further tested by using cobalt nitrate solution as explained under (1) for zinc.

To make this test: Transfer to a flat stick of charcoal a little of the mineral to be tested for arsenic. Use about the amount of the powdered mineral that can be held on the tip of a knife blade or a piece of the mineral about as large as a kernel of wheat. Heat the mineral before the blowpipe in the reducing (yellow) flame. The above-mentioned substances, when thus treated, give off fumes that have a garliclike odor. These arsenical fumes form a white coating on the charcoal, at a distance from the mineral.

Beginners practicing this test should use arsenopyrite.

3. Arsenic and some arsenides when heated in a closed tube with sodium carbonate yield a black sublimate.

To make this test: Mix thoroughly a little of the finely powdered mineral with three volumes of powdered sodium carbonate (baking soda). Place in a closed tube about ½ inch of this mixture. Heat the lower end of the tube at a red heat for some time. Arsenic and some arsenides when given this treatment yield a black, mirrorlike sublimate (coating) on the walls of the tube (arsenical mirror).

Beginners practicing this test should use arsenic.

4. Some orpiments (arsenic trisulfide) fluoresce (glow) green when exposed to strong ultraviolet rays (black light). Always check fluorescent substances by chemical and blowpipe tests.

ASBESTOS

Asbestos is a term applied to several minerals that have flexible, fibrous structures and are more or less acid- and fireproof. Among such minerals are fibrous serpentine (chrysotile) and various amphibole minerals (fibrous tremolite, fibrous actinolite, fibrous anthophyllite, fibrous crocidolite, and fibrous amosite). Chrysotile is the most important asbestos mineral produced in this country.

Asbestos can usually be recognized by its incombustibility, flexible structure, and slow conductivity of heat.

1. To make this test: Twist a few fibers of the material into a string or yarn. Hold one end of the string or yarn in a flame. If the material is asbestos it will not burn.

Beginners practicing this test should compare the combustibility (taking fire and burning) of cotton, wool, and asbestos.

2. The quality of asbestos and its suitability for most uses may be determined by a few simple tests. Length, color, silkiness, flexibility, and, to some extent, fineness of fiber and tensile strength may be determined by inspection. A sample of asbestos should be fiberized by rubbing or crushing between the fingers. Single fibers may then be tested for flexibility and tensile strength by bending and breaking. Several fibers may be twisted into a strand or yarn and again tested for flexibility and strength. Asbestos of good quality should be easily fiberized, soft, silky, strong, flexible, and easily twisted into a strong yarn. Fibers one-fourth inch or more in length and otherwise of a good grade are of commercial interest.[2]

[2]R. B. Ladoo, *Non-Metallic Minerals* (New York: McGraw-Hill Book Co., Inc., 1925), pp. 62, 63.

BARIUM

The most important barium minerals are barite (barytes, heavy spar, or barium sulfate) and witherite (barium carbonate).

1. Volatile compounds of barium (carbonate) color a non-luminous flame yellowish green if heated therein.

To make this test: Use a piece of iron wire (baling wire) about 4 inches long. Wet one end of the wire in dilute (four parts of water to one part of acid) hydrochloric (muriatic) acid so that some of the pulverized mineral will adhere to it. Draw the wet end of the wire through the powdered mineral. Heat the end of the wire with the mineral on it in a flame. An alcohol lamp flame is very satisfactory for this purpose. As soon as the wire and mineral are red hot, the flame will be colored yellowish green if the mineral contains an appreciable amount of a volatile compound of barium. This test can be applied satisfactorily only to barium carbonate (witherite).

2. With a Merwin's Flame-Color Screen: Follow the directions outlined in "Barium" (1) but observe or look at the flame through the different sections of a Merwin's Flame-Color Screen. Through section 1 the barium (or boron) flame is green; through sections 2 and 3 it is a fainter green.

Beginners practicing this test should use witherite.

3. In dilute solutions, dilute sulfuric acid precipitates barium.

To make this test: Place in a test tube or some other glass or porcelain receptacle a little of the mineral to be tested. Use about the amount of the powdered mineral that can be held on the tip of a knife blade. Pour into the receptacle about 4 teaspoonfuls of dilute (equal parts of acid and water) hydrochloric (muriatic) acid. Heat to boiling and then add about 10 teaspoonsfuls of cold water. To this solution add a few drops of dilute (one part acid added to four parts of water) sulfuric acid[3] (oil of vitriol). Upon the addition of the dilute sulfuric acid to the dilute acid solution, any barium in solution will be thrown down as a white precipitate.

Beginners practicing these tests should use witherite.

CALCIUM

Commercially, the most important calcium minerals are calcite (limestone, marble, chalk, or calcium carbonate), dolomite (brown spar or calcium-magnesium carbonate), gypsum (selenite, satin spar, rock gypsum, land plaster, or hydrous calcium sulfate), and fluorite (fluor spar or calcium fluoride).

1. Calcium can usually be detected by its precipitation as an oxalate.

To make this test: Place in a test tube or some other glass or porcelain receptacle a little of the mineral to be tested. Use about

---

[3]Sulfuric acid when mixed with water generates much heat. To make dilute sulfuric acid, always add the acid to the water, a drop or so at a time. *Never* add the water to the acid, as the heat generated may cause an explosion.

the amount of powdered mineral that can be held on the tip of a knife blade. Pour into the receptacle about 1 teaspoonful of concentrated (strong) hydrochloric (muriatic) acid[1] or about twice this amount of dilute (about equal parts of acid and water) hydrochloric acid.

This mixture either cold or on being heated (if the mineral contains calcium carbonate) will effervesce (boil; bubble) and give off a colorless, odorless gas (carbon dioxide) which will not support combustion (a lighted match or other flame if held in this gas will go out). After the effervescence has ceased, add to the liquid about 5 teaspoonfuls of cold water. To this solution add ammonia[2] until a white precipitate begins to form or until the solution smells of ammonia. Then add a few crystals of ammonium oxalate or about ½ teaspoonful of concentrated aammonium oxalate solution (ammonium oxalate crystals dissolved in water). On the addition of the ammonium oxalate to this solution, a white precipitate will be thrown down providing the mineral used in the test contains an appreciable amount of calcium.

Beginners practicing this test should use limestone, marble, chalk, or some other easily soluble calcium mineral.

2. Sulfuric acid precipitates calcium as a sulfate in moderately concentrated solutions.

To make this test: Place in a test tube or some other glass or porcelain receptacle a little of the mineral to be tested. Use about the amount of powdered mineral that can be held on the tip of a knife blade. Pour into the receptacle about 1 teaspoonful of concentrated hydrochloric acid. After all effervescence has ceased, add to this solution a few drops of dilute (one volume of acid added to about four volumes of water) sulfuric acid (oil of vitriol). This precipitates the calcium as colorless, white crystals of calcium sulfate (gypsum) which is distinguished from the sulfates of barium and strontium in that it will dissolve in a solution of ammonium sulfate. Calcium sulfate is also soluble in hot water. If the solution is diluted with water to about ten times its original volume and warmed, the calcium sulfate will dissolve.

Beginners practicing this test should use limestone, marble, chalk, or some other easily soluble calcium mineral.

3. Volatile compounds of calcium with hydrochloric (muriatic) acid color a nonluminous flame yellowish red if heated therein.

To make this test: Wet one end of a piece of iron wire (baling wire) about 4 inches long in hydrochloric (muriatic) acid. Draw the wet end of the wire through some of the powdered mineral. Heat the end of the wire with the mineral on it in a flame. An

---

[1] If barium or strontium is present in the mineral which is being tested, add to the hydrochloric acid solution potassium sulfate or some other alkali sulfate and boil for a few minutes. Filter off the residue and any precipitate that has been formed and proceed with the test.

[2] Ammonia when mixed with an acid generates heat which may cause an explosion. Therefore be certain always to point the end of the receptacle in a direction where no harm can result. Never add ammonia to a *hot* concentrated acid solution.

alcohol lamp flame is very satisfactory for this purpose. As soon as the wire and mineral are red hot, the flame will be colored yellowish red providing the mineral contains an appreciable amount of volatile compounds of calcium and also providing that no other element masks the calcium flame.

Beginners practicing this test should use limestone, marble, chalk, or some other volatile compound of calcium.

4. With a Merwin's Flame-Color Screen: Follow the directions outlined in "Calcium" (3), but look at the flame through the different sections of a Merwin's Flame-Color Screen. Through section 1, calcium gives a flashy, greenish yellow; through section 2, a green; and through section 3, a faint crimson flame.

5. Some calcium minerals (some calcites, limestones, aragonites, gypsums, and others) fluoresce (glow) when exposed to strong ultraviolet light (black light). Red, yellow, orange, and blue are among the colors emitted by these minerals when activated by this light—best seen in the dark.

Always check fluorescent substances by chemical and blowpipe tests.

CALCIUM CARBONATE (LIMESTONE)

Many mining men and prospectors think that a conclusive test for the determination of calcium carbonate (limestone) is that it effervesces (boils; bubbles) when moistened with hydrochloric (muriatic) acid. This effervescing merely indicates that the mineral is a carbonate, providing a colorless, odorless gas which does not support combustion is given off (a lighted match if held in this gas will go out).

Various metals occur as carbonates, a few of the commonest being lead, zinc, copper, and iron. From this statement, it is evident that this effervescence is not a conclusive test for the determination of calcium carbonate, since the sample may contain a carbonate of some other metal. This fallacy has been responsible for failure to appreciate the importance of many valuable mineral deposits.

CHLORINE

1. Chlorine can usually be detected by its precipitation as silver chloride.

To make this test: Place in a test tube or some other glass receptacle a little of the material to be tested. If the material is a mineral use about the same amount of the powdered mineral that can be held on the tip of a knife blade; if the material is a liquid use about 1 teaspoonful. Pour into the receptacle about 1 teaspoonful of dilute nitric acid. Heat this mixture to boiling and then cool it. On the addition of a small amount of silver nitrate to this cool solution a white precipitate will be thrown down providing the material used in the test contains an appreciable amount of chlorine. If only a small amount of chlorine is present in the material tested, the precipitate gives the solution a milky appearance; if considerable chlorine is present, the precipitate looks curdy. If exposed to the sunlight for a time, this white precipitate turns dark (from a violet to brown color). This precipitate (silver chloride) is also soluble in ammonia. To make this part of the test, pour into the receptacle an excess of ammonia (until the solution smells strong of ammonia). The ammonia will dissolve the silver chloride.

2. Chlorine can usually be detected by the fumes that are given off when a chloride is heated with potassium bisulfate and manganese dioxide. These fumes have a bleaching action.

To make this test: Mix thoroughly a little of the finely powdered mineral with an equal volume of manganese dioxide (pyrolusite, psilomelane, or wad) and about four volumes of powdered potassium bisulfate (acid sulfate of potassium). Fill a small test tube to about ½ inch from the bottom with this mixture. Heat the lower end of the tube at a red heat for several minutes. This mixture when given this treatment will give off rusty-green fumes that have a strong pungent odor providing the material used in the test contains an appreciable amount of chlorine. These rusty-green fumes, if chlorine gas, have a bleaching action. This can be tested by holding a strip of moistened litmus paper inside the tube in the fumes given off.

Beginners practicing these tests should use halite (sodium chloride, or common table salt) or sylvite (potassium chloride).

CHROMIUM

The most important chromium mineral is chromite (chromic iron ore or ferrous metachromite).

1. Chromium can usually be detected by the colors it imparts to the fluxes.

To make this test: Use a piece of No. 26 platinum wire about 2 inches long. Fasten one end in a holder so that when the wire is heated it will not burn the fingers. Special holders can be purchased, but a cork or a piece of soft wood into which one end of the wire is inserted may be used for this purpose. Make a small loop about 1/16 inch in diameter at the unattached end of the wire. This loop is easily made by winding the end of the wire around the point of a lead pencil. Heat this looped end in a flame until it is red hot. An alcohol lamp flame is very satisfactory for this work. Dip the red-hot loop into some powdered borax or salt of phosphorus, a little of which will adhere to the wire loop. Fuse the borax or salt of phosphorus adhering to the wire by holding it in the flame. Continue these operations until a clear, glassy bead that fills the loop in the wire is secured. Touch the bead while it is red hot to a little of the very finely powdered mineral. If the bead[6] made from the borax and a very little of a mineral

---

[6]The chromium bead tests must not be confused with those for vanadium, which give in the reducing flame almost identical reactions with the fluxes, but vanadium in the oxidizing (bluish) flame differs from the salt of phosphorus bead test in that it yields a yellow bead, while this flux never yields other than a green bead with chromium.

containing chromium is heated before the blowpipe in the oxidizing (bluish) flame, the bead will be decidedly yellow while it is warm, changing to a yellowish-green color when cold. When more of the mineral is added, the colors are deeper, changing through reddish or yellow when warm to a fine green when cold. If this bead is heated before the blowpipe in the reducing (yellow) flame, the bead assumes a fine, green color when cold, but shows none of the yellow or reddish tint which is so prominent in the warm bead after heating in the oxidizing flame.

If salt of phosphorus is used instead of borax for making the bead, and the operations outlined above for the borax bead test are followed, the color of the bead formed when the oxidizing flame of the blowpipe is employed is a dirty green when the bead is warm, which changes to a fine green when the bead is cold. If the reducing blowpipe flame is used. the colors are about the same as with the oxidizing flame.

Beginners practicing the chromium test should use chromite.

2. Chromium can usually be detected by its yellow colored sulfuric acid solution. This turns green on the addition of alcohol.

To make this test: Mix one part of powdered ore with ten parts of powdered manganese dioxide. Place a little of this mixture in a test tube or some other glass or porcelain receptacle. Use about three times the amount of this mixture that can be held on the tip of a knife blade. Pour into the receptacle about 1 teaspoonful of water and the same amount of concentrated (strong) sulfuric acid (oil of vitriol). Heat this solution to boiling and boil strongly until dense white fumes are freely given off. Then cool the solution and, after it is cold, dilute it with about 3 teaspoonfuls of cold water. Filter the solution and catch the filtrate (the clear liquid that is filtered and passes through the filter paper) in a glass receptacle. This filtrate will be yellow providing the mineral used in the test contained an appreciable amount of chromium.

About 1 teaspoonful of alcohol is then added to this yellow colored solution which is then boiled. The solution will slowly turn green, due to the reduction of the chromium to chromic sulfate, providing the mineral contained an appreciable amount of that metal.

COBALT

The most important cobalt minerals are smaltite (tin-white cobalt or cobalt-nickel arsenide) and cobaltite (cobalt arsenide-sulfide).

1. Cobalt can usually be detected by the blue color it imparts to the fluxes.

To make this test: Use a piece of No. 26 platinum wire about 2 inches long. Fasten one end in a holder so that when the wire is heated it will not burn the fingers. Special holders can be purchased, but a cork or piece of soft wood may be used for this purpose. Make a loop about 1/16 inch in diameter at the unattached end of the wire. This loop is easily made by winding the end of the wire around the point of a lead pencil. Heat this looped end in a flame until it is red hot. An alcohol lamp is very satisfactory for this work. Dip the red-hot loop into some powdered borax or salt of phosphorus, a little of which will adhere to the wire loop. Fuse the borax or salt of phosphorus adhering to the wire by holding it in the flame. Continue these operations until a clear, glassy bead that fills the loop in the wire is secured. Touch the bead while it is red hot to a very little of the finely powdered mineral.[7] If the beads made from either borax or salt of phosphorus and a cobalt mineral are heated before the blowpipe in either the reducing (yellow) flame or the oxidizing (bluish) flame, the color of the beads formed will be deep blue.

Beginners practicing this cobalt test should use roasted smaltite or roasted cobaltite.

2. Cobalt compounds become magnetic when heated on charcoal before the blowpipe in the reducing flame.

To make this test: Mix thoroughly a little of the finely powdered mineral with about twice its volume of powdered sodium carbonate (baking soda). Transfer to a stick of charcoal as much of this mixture as can be held on the tip of a knife blade. Heat strongly before the blowpipe in the reducing flame until the mixture is thoroughly fused. The resulting fused mass will contain a dark colored, more or less metallic button enclosed in the slag, and this button[8] will be magnetic when cold if the mineral used contains cobalt.

Beginners practicing this test should use smaltite or cobaltite.

COPPER

Important copper minerals are native copper, chalcocite (copper glance, vitreous copper, or cuprous sulfide), chalcopyrite (copper pyrites, yellow copper ore, fool's gold, or sulfide of copper and iron), bornite (purple copper ore, variegated copper ore, horse-flesh ore, peacock copper, or sulfide of copper and iron), malachite (green copper carbonate or basic carbonate of copper), azurite (blue copper carbonate or basic carbonate of copper), cuprite (ruby copper, red copper ore, or cuprous oxide), and copper-bearing pyrites.

1. Ammonia added to an acid solution of copper produces a blue coloration.

To make this test: Place in a test tube or some other glass or porcelain receptacle a little of the mineral to be tested. Use about the amount of powdered mineral that can be held on the tip of a

---

[7] A sulfide or arsenide ore must be thoroughly roasted (pulverized and heated on charcoal at a red heat until sulfur or arsenic fumes are no longer noticeable) before using the cobalt mineral in the bead tests.

[8] Metallic iron and nickel are also magnetic; therefore a magnetic button obtained from any mineral, the metallic contents of which are unknown, should be tested further by applying test (1) for cobalt.

knife blade. Pour into the receptacle about 1 teaspoonful of acid. Use concentrated (strong) nitric acid or a mixture of nitric and hydrochloric (muriatic) acid. Heat until the copper has been dissolved[9] and then add about 3 teaspoonfuls of cold water. To this add an excess of ammonia[10] (until the solution smells strong of ammonia). Upon the addition of the ammonia to the acid solution, the color of the solution will turn blue if the mineral used in the test contains an appreciable amount of copper.

Beginners practicing this test should use malachite, azurite, chalcopyrite, chalcocite, or some other high grade copper mineral or a small piece of metallic copper.

2. Clean iron, steel, zinc, or aluminum, if immersed in a dilute acid solution of copper, will become coated (plated) with a film of copper.

3. Volatile compounds of copper color a nonluminous flame green if heated therein. With hydrochloric acid the flame is colored azure-blue.

To make this test: Use a piece of iron wire (baling wire) about 4 inches long. Wet one end of the wire in water so that some of the pulverized mineral will adhere to it. Draw the went end of the wire through the powdered mineral. Heat the end of the wire with the mineral on it in a flame. An alcohol lamp flame is very satisfactory for this purpose. As soon as the wire and mineral are red hot, the flame will be colored greenish if the mineral contains an appreciable amount of a volatile compound of copper.

If the same end of the wire[11] is moistened with hydrochloric acid, more of the mineral taken upon it, and it is again held in the flame, the flame will be colored azure-blue if the mineral contains an appreciable amount of a volatile compound of copper.

Beginners practicing this test should use malachite (copper carbonate) or some other volatile copper mineral.

4. With a Merwin's Flame-Color Screen: Follow the directions outlined in "Copper" (3) but observe or look at the flame through the different sections of a Merwin's Flame-Color Screen. Through

---

[9]If the mineral does not dissolve readily fuse it as directed in "Copper" (5) and use the fused mass in this test.

[10]Ammonia when mixed with an acid generates heat. This heat may cause an explosion. Therefore be certain always to point the end of the receptacle in a direction where no harm can result. Never add ammonia to a *hot* acid solution.

In order not to crack glass receptacles, heat very gently at first, shaking the receptacle so that the solution washes around the bottom and sides.

[11]Never dip a wire which has been used for this or other tests in the acid bottle, since any soluble mineral present will dissolve and the acid may be made valueless for further flame tests. Pour a few drops of the acid into a glass or cup, and dip the end of the wire into it.

The copper of volatile copper compounds alloys with the platinum when platinum wire is used for making flame tests, making the alloyed piece of wire worthless for further copper flame tests.

When iron wire is used, use a new piece of wire for each copper flame test.

section 1, copper gives a bright green; through section 3, a bright blue flame fringed with green; and through section 2, the same tints, but paler.

5. Copper compounds fused with soda in the reducing flame yield a mass of metallic copper.

To make this test: Mix thoroughly a little of the finely powdered mineral with about twice its volume of powdered flux (sodium carbonate [baking soda]—alone or mixed with a little borax). Transfer to a stick of charcoal as much of this mixture as can be held on the tip of a knife blade. Heat this mixture strongly before the blowpipe in the reducing (yellow) flame until it is thoroughly fused. The resulting fused mass will contain an irregular, spongy mass of metallic copper if the mineral used in the test contains an appreciable amount of copper.

Beginners practicing this test should use chalcopyrite, cuprite, or some other easily fusible copper mineral.

6. Copper can usually be detected by the colors it imparts to the fluxes. Borax and salt of phosphorus beads are green when warm and blue when cold in the oxidizing (bluish) flame. Saturated beads are opaque red in the reducing flame of the blowpipe. In the presence of much iron, the oxidizing flame bead is green or bluish green. Bead test instructions can be found under "Chromium" or "Cobalt."

7. Some copper minerals (some azurites, chrysocollas, and malachites) fluoresce (glow) blue, gray, or white when exposed to strong ultraviolet rays (black light).

Always check fluorescent substances by chemical and blowpipe tests.

FLUORINE

The most important fluorine minerals are fluorite (fluor spar or calcium fluoride) and fluorapatite.

1. Fluorine etches hard glass.

a) Some compounds of fluorine with sulfuric acid etch hard glass.

To make this test: Take a flat piece of window glass and coat a section of one side of the glass with a thin, even layer of paraffin wax. Wax melted and dropped from a burning paraffin candle can be used for this purpose. Allow the wax to cool and when cold write or mark through the paraffin to the glass. Use a pencil point, a sharpened piece of wood, or some other tool that will cut through the paraffin, but will not scratch the glass. Pour into the marks made in the paraffin some of the finely powdered mineral that is being tested for fluorine. Onto this powdered mineral pour a few drops of concentrated (strong) sulfuric acid (oil of vitriol), and mix the mineral and acid together to the consistency of a thick paste. Use a pointed stick or pencil point for this purpose. In about 10 minutes wash off the mixture of acid and mineral, remove the paraffin by heating the glass, and clean the glass. Etching of the glass indicates that fluorine was present in the mineral used in making this test.

Beginners practicing this test should use fluorite (fluor spar).

b) With potassium bisulfate, some compounds of fluorine etch glass.

To make this test: Mix thoroughly a little of the finely powdered mineral with about three volumes of potassium bisulfate (acid sulfate of potassium). Fill a closed tube or a small test tube to about ½ inch from the bottom end with this mixture. Heat the lower end of the tube at a red heat for several minutes. Etching (roughening or clouding) of the sides of the tubes, just above the charge, indicates that fluorine is present in the mineral tested. The etching may be detected by breaking the tube, washing a fragment thoroughly, and rubbing the surface with a sharp point of a knife blade. The glass will feel slightly rough if it has been etched.

Beginners practicing this test should use fluorite.

2. a) Some varieties of fluorite become phosphorescent when heated.

To make this test: Place in a test tube a few fragments of the mineral to be tested. The fragments should be about ¼ inch in size. Heat the lower portion of the tube at a red heat, for a very short time. If heated too long the glowing will disappear. Some varieties of fluorite, when thus treated and held in the dark, become phosphorescent (glow and emit light of various tints).

Beginners practicing this test should use violet-colored fluorite.

b) The above test can also be applied to a splinter of the material held in a pair of forceps.

3. Some fluorites (calcium fluoride) fluoresce (glow) a blue or green color when exposed to strong ultraviolet light (black light).

Always check fluorescent substances by chemical and blowpipe tests.

GOLD

The principal gold minerals are native gold, sylvanite (gold and silver telluride), and calaverite (gold telluride).

1. The following charcteristics serve for the ready detection of gold: Its brass-yellow color; its high specific gravity (heaviness); its high fusibility (the high temperature required to melt it); gold boils at about 4,700 degrees Fahrenheit; its malleability (gold can be flattened out if hammered on an anvil; a knife blade, needle, or similar tool cuts and indents gold without crushing, cracking, or breaking it, as with metallic lead); and its insolubility (nitric acid alone, hydrochloric acid alone, or sulfuric acid alone does not dissolve gold). Gold, however, is soluble in aqua regia (nitric acid mixed with hydrochloric acid).

2. **Amalgamation.**[12]—Amalgamation is the process of uniting mercury (quicksilver) with another metal. Amalgamation as used in this test is based upon the fact that when *clean, bright* gold is brought into contact with *clean, bright* mercury, especially by a *rubbing or grinding action*, the mercury sticks to, coats, and catches the gold. When particles of mercury-coated gold come in contact with each other, they become loosely cemented or soldered together. The resulting mass or paste is gold amalgam.

If the mercury is dark or tarnished, the gold, no matter how bright and clean it is, will not be caught by or unite with the mercury. Neither will the union take place if the gold is rusty or dirty even though the mercury is bright and clean. Both the gold and the mercury must be bright and clean to unite. Grinding the mixture in cyanide solution (very poisonous) brightens the gold and cleans the mercury.

If gold amalgam is heated before the blowpipe on charcoal, the mercury will be distilled and leave the gold as a residue. If this residue, mixed with a little powdered borax, is then heated before the blowpipe on charcoal, there will be obtained a malleable, brass-yellow button which can be tested as explained in "Gold" (1).

3. **Panning.**—Gold can usually be detected in free-milling ores, sand and gravel by panning.

With many complex, refractory (rebellious) ores, panning does not give satisfactory results. For such, a preliminary roasting of the ore often overcomes the difficulty.

When the above tests do not indicate the presence of gold but the operator still believes that the mineral contains gold, the policy recommended for testing such a mineral is to have it assayed for gold. If an assay does not reveal the presence of gold in a mineral, it can be safely assumed that gold is not present in that mineral in commercial quantity.

4. In nitrohydrochloric acid solutions of gold, stannous chloride[13] gives a purple precipitate. Metallic tin gives the same coloration.

To make this test: Place in a test tube or some other glass or porcelain receptacle a little of the powdered mineral to be tested. Use about the amount of powdered mineral that can be held on the tip of a knife blade. Pour into the receptacle about 1 teaspoonful of concentrated (strong) nitric acid and 4 teaspoonfuls of concentrated hydrochloric (muriatic) acid. Heat this solution to boiling and boil until any gold present has been dissolved. To this solution add a pinch of stannous chloride or a pinch of metallic tin filings. Upon the addition of an excess of either of

---

[12]Fuller information on amalgamation is given in *Arizona Gold Placers and Placering* (Univ. of Ariz., Ariz. Bur. Mines Bull. No. 135).
[13]These salts are rather unstable. If stannous chloride changes to stannic chloride it is worthless for this test. The same is true if the ferrous sulfate changes to ferric sulfate.

these reagents to the nitrohydrochloric acid, the solution will turn a deep purple color providing the mineral used in the test contained an appreciable amount of gold. On exposure to the air the purple solution turns yellow. This test is known as the "Purple of Cassius Test for Gold."

5. In nitrohydrochloric acid solutions of gold, ferrous sulfate gives a brown precipitate.

To make this test follow the directions outlined in "Gold" (4) but use ferrous sulfate instead of stannous chloride.

Beginners practicing these tests should use a piece of metallic gold or concentrates from a gold ore.

IRON

Important iron minerals of commerce are hematite (red ocher, red oxide of iron, specular iron, iron glance, ferric oxide, or iron sesquioxide), limonite (brown oxide of iron, brown ocher, brown hematite, or bog iron ore), and magnetite (black iron oxide, magnetic iron ore, or lodestone).

1. Iron minerals roasted on charcoal yield a magnetic residue.

To make this test: Place on a stick of charcoal a little of the finely powdered mineral to be tested. Use about the amount that can be held on the tip of a knife blade. Heat before the blowpipe in the reducing (yellow) flame. If the mineral thus heated *without a flux* does not fuse (melt), but becomes magnetic, it contains iron. If it fuses and becomes magnetic it may contain iron, cobalt, or nickel.

Beginners practicing this test should use hematite, pyrites, or some other iron mineral.

2. Iron minerals fused with soda in the reducing flame yield a magnetic button.

To make this test: Mix thoroughly a little of the finely powdered mineral with about twice its volume of sodium carbonate (baking soda). Transfer to a stick of charcoal as much of this mixture as can be held on the tip of a knife blade. Heat strongly before the blowpipe in the reducing flame until it is thoroughly fused. The resulting fused mass will contain a dark-colored, more or less metallic button[14] which is magnetic when cold, providing the mineral used in the test contains an appreciable amount of iron.

Beginners practicing this test should use hematite, pyrites, or some other easily fusible iron mineral.

3. Ammonia added to an acid solution of iron throws down a brownish-red precipitate.

To make this test: Place in a test tube or some other glass or porcelain receptacle a little of the mineral to be tested. Use about the amount of powdered mineral that can be held on the tip of a knife blade. Pour into the receptacle about 1 teaspoonful of

---

[14]Cobalt and nickel buttons produced in this manner are also magnetic. For that reason the magnetic button should be further tested for iron by "Iron" (3) and for cobalt and nickel by the test given for those metals.

concentrated (strong) hydrochloric (muriatic) acid and a few drops of nitric acid. Heat this solution until the iron has been dissolved,[15] and then add about 2 teaspoonfuls of cold water. To this solution add an excess of ammonia[16] (until the solution smells strong of ammonia). Upon the addition of the ammonia to this acid solution, there will be thrown down a brownish-red precipitate if the mineral used in the test contains an appreciable amount of iron.

Beginners practicing this test should use hematite, limonite, or some other easily soluble iron mineral or a small piece of metallic iron.

4. Iron can usually be detected by the colors it imparts to the fluxes. In the oxidizing (bluish) flame of the blowpipe, the borax bead of iron is amber-colored when warm and yellow to colorless when cold, while in the reducing flame the borax bead is bottle-green, providing the bead is saturated. Bead test instructions are given under "Chromium" and "Cobalt."

5. Some limonites (brown hematite or iron sesquioxide) and some siderites (iron protocarbonate) fluoresce (glow) green when exposed to strong ultraviolet rays (black light).

Always check fluorescent substances by chemical and blowpipe tests.

LEAD

The most important lead minerals of commerce are galena (galenite, steel galena, potter's ore, or lead sulfide), cerussite (lead carbonate or white lead ore), and anglesite (lead sulfate).

Lead can usually be detected by the yellow sublimate and the metallic lead button formed on charcoal.

To make this test: Mix thoroughly a little of the finely powdered mineral with an equal volume of powdered charcoal and three volumes of powdered sodium carbonate (baking soda). Moisten this mixture with water and transfer about the amount that can be held on the tip of a knife blade to a flat piece of charcoal or into a shallow cavity that has been made in the charcoal. Heat this before the blowpipe in a moderately strong, reducing (yellow) flame. This treatment, if the mineral used in the test contains an appreciable amount of lead, will produce small globules or buttons of metallic lead, which are soft and malleable (can be flattened out if hammered on an anvil), and also a yellowish sublimate (coating) on the charcoal, close to the mineral. This sublimate is whitish on the outer edges, and the white portion should not be confused with an antimony or zinc sublimate.

---

[15]Many iron minerals are practically insoluble in the acids just mentioned. These minerals if roasted or treated as described in "Iron" (2) become easily soluble.

[16]Ammonia when mixed with an acid generates heat. This heat may cause an explosion. Therefore be certain always to point the end of the receptacle in a direction where no harm will result. Never add ammonia to a *hot* acid solution.

Beginners practicing this test should use galena, cerussite, or some other high-grade lead mineral or some small cuttings of metallic lead.

2. Hydrochloric or sulfuric acid throws down a heavy, white precipitate in cold, nitric acid solutions of lead.

To make this test: Place in a test tube or some other glass or porcelain receptacle a little of the mineral to be tested. Use about the amount of the powdered mineral that can be held on the tip of a knife blade. Pour into the receptacle about 1 teaspoonful of concentrated (strong) nitric acid and about 2 teaspoonfuls of water. Boil this solution until the lead has been dissolved, and then cool to room temperature. When cold add a few drops of dilute sulfuric acid (oil of vitriol) or a few drops of hydrochloric (muriatic) acid.[17] Upon the addition of either of these acids to the cold nitric acid solution there will appear a white, heavy precipitate of lead providing the mineral used in the test contains any appreciable amount of lead. This white precipitate should be furthur tested for lead by using test "Lead" (1).

Beginners practicing this test should use cerussite, or some other easily soluble, high-grade lead mineral, the lead buttons produced in test "Lead" (1), or some cuttings of metallic lead.

The white precipitate (lead chloride) formed when hydrochloric acid is added to a nitric acid solution of lead can be tested further as follows:

3. Lead chloride is quite soluble in hot water.

To make this test: Add from ten to fifteen volumes of water to the mixture obtained when hydrochloric acid is used in making test "Lead" (2). Heat to boiling and boil for a minute or two. If the white precipitate from test "Lead" (2) is lead chloride, it will dissolve.

4. Lead sulfate (anglesite) gives cracking sounds and decrepitates (flies to pieces) before a hot flame. Use an alcohol torch flame or a carbide lamp flame for this work.

Lead carbonate (cerussite) is commonly associated with lead sulfate. Lead carbonate like lead sulfate fuses easily. In a closed tube lead carbonate decrepitates.

5. Some cerussites (lead carbonate) fluoresce (glow) a pale blue color when exposed to strong ultraviolet light (black light).

Always check fluorescent substances by chemical and blowpipe tests.

MAGNESIUM

The most important magnesium mineral is magnesite (magnesium carbonate).

---

[17]When hydrochloric acid is listed in test "Lead" (2) silver and mercury, if present, will also be thrown down as a white precipitate, but silver in this form (silver chloride) turns dark if exposed to sunlight and is also very soluble in ammonia. The white mercury precipitate thus obtained usually turns dark grayish on the addition of an excess of ammonia.

1. Magnesium can usually be detected by its precipitation as ammonium-magnesium phosphate.

To make this test: Place in a test tube or some other glass or porcelain receptacle a little of the mineral to be tested. Use about the amount of powdered mineral that can be held on the tip of a knife blade. Pour into the receptacle about 3 teaspoonfuls of hydrochloric (muriatic) acid, a drop of nitric acid and about 4 teaspoonfuls of water. Heat to boiling, and then cool to room temperature. To the cold solution add an excess of ammonia (until the solution smells strong of ammonia). If a precipitate is thrown down, filter the precipitate off and catch the clear, filtered solution in another glass receptacle. To this clear liquid add a little ammonium carbonate or a little ammonium oxalate solution. Again filter off any precipitate which may form, and catch the clear, filtered solution in another receptacle. To this last clear solution add a few drops of sodium phosphate solution. This will cause the formation of a white, crystalline precipitate of ammonium-magnesium phosphate[18] providing the mineral used in this test contains an appreciable amount of magnesium.

Beginners practicing this test should use magnesite, dolomite, or some other easily soluble magnesium mineral.

2. Some of the white or colorless, magnesium compounds, such as magnesite, when moistened with cobalt nitrate and heated before the blowpipe, assume a light pink, or flesh color. The following method for making this test is given in G. M. Butler's *Handbook of Blowpipe Analysis*.

To make this test:

Hold a small splinter of the substance to be tested in the platinum forceps and heat it in the blowpipe flame to the higest possible temperature. Then examine it with a lens; if it shows any signs of fusion, this test cannot be applied. If non-fusible, moisten it with cobalt nitrate and ignite strongly in the hottest part of the blowpipe flame. It will first turn black, but after prolonged heating may assume a characteristic tint. If a splinter of the substance cannot be obtained, it should be powdered and the test conducted upon a flat cake of the powder upon charcoal. Longer heating is required by this method however, and the results are not apt to be as satisfactory.

This test can be applied only to nonfusible, white or faintly tinted minerals, or those which become white or faintly tinted upon ignition.

A pinkish or flesh-tinted coloration indicates magnesium.

Beginners practicing this test should use magnesite.

3. Some dolomites (carbonate of calcium and magnesium) fluoresce (glow) a white or gray color when exposed to strong ultraviolet light (black light).

Always check fluorescent substances by chemical and blowpipe tests.

---

[18]In order to allow sufficient time for the ammonium-magnesium phosphate precipitate to form, it is sometimes necessary to let the solution stand up to 12 hours.

MANGANESE

Important manganese minerals are pyrolusite (black oxide of manganese, or manganese dioxide), psilomelane (impure hydrous manganese dioxide), rhodochrosite (manganese protocarbonate), rhodonite (manganese metasilicate), and bementite (hydrous manganese silicate).

1. Manganese can usually be detected by the colors it imparts to the fluxes.

a) To make this test: Use a piece of No. 26 platinum wire about 2 inches long. Fasten one end in a holder so that when the wire is heated it will not burn the fingers. Special holders can be purchased, but a cork or piece of soft wood may be used for this purpose. Make a small loop about 1/16 inch in diameter at the unattached end of the wire. This loop is easily made by winding the end of the wire around the point of a lead pencil. Heat this looped end in a flame until it is red hot. An alcohol lamp flame is very satisfactory for this work. Dip the red-hot loop into some powdered borax, a little of which will adhere to the wire loop. Fuse the borax adhering to the wire by holding it in the flame. Continue these operations until a clear, glassy bead that fills the loop in the wire is secured. Touch the bead while it is red hot to a little of the very finely powdered mineral. If the bead made from the borax and a *very little* manganese mineral[19] are heated before the blowpipe in the oxidizing (bluish) flame, there is produced a bead which is opaque while it is hot, but on cooling it becomes transparent and has a fine, reddish-violet or amethystine color. If this manganese borax bead is heated for a time in the reducing (yellow) flame of the blowpipe, it becomes colorless when cold. The bead test is a very delicate one for manganese.

Beginners practicing this test should use pyrolusite, psilomelane, or some other oxidized manganese mineral.

b) If sodium carbonate (baking soda) is used instead of borax for making the bead test, and the operations just described for conducting the borax bead test "Manganese" (1) (a) are followed, the color of the bead which is formed when the oxidizing flame of the blowpipe is employed is green when the bead is warm and greenish blue when it cools. The sodium carbonate bead made in the reducing flame is colorless. Sodium carbonate beads are opaque.

If the greenish-blue sodium carbonate bead of manganese is

---

[19]Use only a very little of the mineral in this test because if too much is added to the bead, the color produced will be so intense that it will appear to be black, thus making the test valueless.

A sulfide or arsenide ore must be thoroughly roasted (pulverized and heated on charcoal at a red heat until sulfur or arsenic fumes are no longer noticeable) before using the mineral in the bead test.

If much iron is present in the mineral tested, the borax bead will have a reddish-brown color.

dissolved in a drop of water on a piece of glass, and to this solution is added a drop of nitric acid, the solution will become pink.

Beginners practicing this test should use pyrolusite, psilomelane, wad, or some other oxidized manganese mineral.

2. Nitric acid solutions of manganese with sodium bismuthate give a wine-purple (permanganate) colored solution.

To make this test: Place in a test tube or some other glass or porcelain receptacle a little of the mineral to be tested. Use about the amount of the powdered mineral that can be held on the tip of a knife blade. Pour into the receptacle about 2 teaspoonfuls of concentrated (strong) nitric acid and about 1 teaspoonful of water. Boil the mixture for 1 or 2 minutes and then let it stand for a time so that all residue settles.

In a second test tube place a pinch of sodium bismuthate powder. Pour onto the sodium bismuthate about 1 inch of the clear nitric acid solution from the first receptacle. Then add some cold water and let the mixture stand until all residue has settled. Manganese, if present in the mineral used in the test gives a wine-purple (permanganate) colored solution with sodium bismuthate, providing an excess of chlorides or other reducing substances is not present.

Beginners practicing this test should use pyrolusite, psilomelane, wad, or some other oxidized manganese mineral.

3. Red oxide of lead when added to nitric acid solutions of manganese usually gives a pink (permanganate) colored solution.

To make this test: Place in a test tube or some other glass or porcelain receptacle a little of the mineral to be tested. Use about the amount of powdered mineral that can be held on the tip of a knife blade. Pour into the receptacle about 1 teaspoonful of concentrated nitric acid and about the same amount of water. Heat the mixture to boiling and boil for a minute or two and then let it stand for a time, so that all residue settles.

In a second test tube place a pinch of powdered red oxide of lead (red colored litharge can be use). Pour onto the red oxide of lead about 1 inch of the clear nitric acid solution from the first receptacle. Then add a little cold water and let the mixture stand until all residue has settled. Manganese, if present in the mineral used in the test, gives a pink (permanganate) colored solution with red oxide of lead, providing an excess of chlorides or other reducing substances is not present.

Beginners practicing this test should use pyrolusite, psilomelane, or some other oxidized manganese mineral.

4. Many manganese minerals dissolve in hydrochloric acid with the evolution of chlorine gas.

To make this test: Place in a test tube or some other glass or porcelain receptacle a little of the mineral to be tested. Use about the amount of the powdered mineral that can be held on the tip of a knife blade. Pour into the receptacle about 1 teaspoonful of concentrated hydrochloric (muriatic) acid. Heat gently at first, then increase the temperature as much as possible, and chlorine gas will be given off providing the mineral used in the test contains an appreciable amount of manganese oxides. This gas can be recognized by its yellowish-green color and by its pungent odor.

Beginners practicing this test should use pyrolusite, psilomelane, or some other oxidized manganese mineral.

5. Some manganese minerals when heated yield oxygen.

To make this test: Pour into a closed tube a few fragments or a small amount of a powdered manganese dioxide mineral such as pyrolusite or psilomelane. Place a sliver of charcoal in the tube a little above the mineral. Heat the tube so that the charcoal alone is heated, and it will be noticed that although the charcoal gets red hot it does not burn, owing to the limited supply of oxygen in the tube. Keeping the charcoal red hot, apply the heat to the tube so that both the manganese dioxide mineral and the charcoal are heated. As soon as oxygen commences to be given off from the mineral the charcoal will burn brightly.

6. Some rhodochrosites (manganese protocarbonate) fluoresce (glow) a gray color when exposed to strong ultraviolet rays (black light).

Always check fluorescent substances by chemical and blowpipe tests.

MERCURY (QUICKSILVER)

The most important mercury mineral is cinnabar (native vermilion or mercuric sulfide).

1. Mercury when heated with soda in a closed tube condenses as metallic globules of mercury on the sides of the tube.

To make this test: Mix thoroughly a little of the finely powdered mineral with about three volumes of *dry* sodium carbonate (baking soda). Place in a closed tube about ½ inch of this mixture and cover this layer with an additional layer of soda to a depth of about ½ inch. Heat carefully and mercury will distill and condense as globules on the walls of the tube providing the mineral used in the test contains an appreciable amount of mercury (quicksilver). If only a little mercury is formed, it will appear as a gray sublimate (coating) composed of minute globules which may be made to unite by rubbing with a splinter of wood.

Beginners practicing this test should use cinnabar or metallic mercury.

2. a) Most compounds of mercury, if moistened with hydrochloric (muriatic) acid and rubbed on a piece of bright copper, will coat the copper. The copper will then appear as if it had been silver plated.

In this test, quicker results are usually obtained if the mineral is powdered instead of in a chunk. The addition of a little powdered manganese dioxide speeds up the reaction.

b) This precipitation test can also be performed by boiling the mineral with hydrochloric acid in a test tube or some other glass or porcelain receptacle. Addition of a little powdered manganese dioxide expedites the reaction. A piece of bright copper immersed in this solution becomes covered by a thin coating of metallic mercury, providing the mineral used in the test contains an appreciable amount of mercury.

Beginners practicing this test should use cinnabar or metallic mercury.

3. W. G. Leighton (1935) has described a method by which small amounts of mercury may be detected in a sample of ore by an indirect fluorescent means. The sample to be tested is powdered and heated over a Bunsen burner or similar source of heat. During heating of the sample an ultraviolet light source is placed close to the sample, and a willemite-coated screen is placed behind the sample. The mercury in the sample is volatilized and causes a dense shadow to appear on the willemite screen. Without the presence of volatilized mercury the willemite screen will fluoresce a uniform strong green over the entire surface. Ordinary smoke has little effect upon the willemite screen, but mercury vapor appears as a dense, black cloud of smoke against the screen.[20]

MOLYBDENUM

The most important molybdenum minerals of commerce are molybdenite (molybdenum sulfide or molybdenum disulfide) and wulfenite (lead molybdate).

1. In acid solutions of molybdenum, sodium or potassium ethyl xanthate usually gives a pink to purple precipitate.

To make this test: Place in a large test tube or some other glass or porcelain receptacle about ¼ teaspoonful of the powdered mineral to be tested (the receptacle should hold at least 100 cubic centimeters, approximately 7 tablespoonfuls). Pour onto the mineral about 1 teaspoonful of hydrochloric (muriatic) acid, about 1 teaspoonful of nitric acid and about 1 teaspoonful of sulfuric (oil of vitriol) acid. Heat this mixture to boiling and boil it until dense white fumes are given off freely. Add about 2 teaspoonfuls of water and boil the mixture again for a minute or so. To this add strong potassium or sodium hydroxide[21] solution until the solution is slightly alkaline (alkalies turn red litmus blue), after which, boil it for a minute or two.

Let the mixture cool and settle; then filter out the precipitate (this contains the interfering substances), catching the filtrate (the clear solution that passes through the filter) in another test tube or glass receptacle. Pour about 1 inch of this filtrate into a 6 inch test tube and acidify (acids turn blue litmus red) it by adding an excess of hydrochloric or sulfuric acid. Pour into

[20]H. C. Dake and Jack DeMent, *Fluorescent Light and Its Applications* (Brooklyn, N.Y., Chemical Publishing Company, Inc., 1941), pp. 219.
[21]In order to reduce sliminess, some add a large pinch of sodium carbonate (baking soda) or about 2 teaspoonfuls of ammonia, before the hydroxide solution is added.

another 6 inch test tube about 1 inch of strong sodium or potassium ethyl xanthate solution (the salt dissolved in water). Pour some of the acidified filtrate into the xanthate[22] solution.

If molybdenum is present in the mineral used in the test, a pink to purple precipitate (sometimes fading out rapidly) is usually given when the acidified filtrate mixes with the sodium or potassium xanthate solution.

Beginners practicing this test should use molybdenite, wulfenite, or some other molybdenum mineral.

2. In nitric acid solutions of molybdenum potassium ferrocyanide throws down a reddish-brown precipitate.

To make this test: Place in a test tube or some other glass or porcelain receptacle a little of the mineral to be tested. Use about the amount of powdered mineral that can be held on the tip of a knife blade. Pour into the receptacle about 3 teaspoonfuls of concentrated (strong) nitric acid. Boil for a few minutes and then add 4 or 5 teaspoonfuls of cold water. Upon the addition of ferrocyanide to the acid solution, a reddish-brown precipitate will be thrown down providing the mineral used in the test contains an appreciable amount of molybdenum.

Beginners practicing this test should use molybdenite or wulfenite.

3. Molybdenum can usually be detected by the colors it imparts to the fluxes.

a) In the oxidizing (bluish) flame of the blowpipe, borax beads[23] of molybdenum are yellow when warm and colorless when cold. In the reducing (yellow) flame the warm and cold borax beads are colorless. These beads made in the reducing flame when saturated with molybdenum are brown when warm or cold.

b) All salt of phosphorus beads of molybdenum are green except the unsaturated, cold bead formed in the oxidizing (bluish) flame which is colorless.

c) If several of the green salt of phosphorus beads are dissolved in dilute hydrochloric acid and tin is added to the boiling solution, the solution turns brown.

4. Some wulfenites (lead molybdate) fluoresce (glow) a green color when exposed to strong ultraviolet rays (black light).

Always check fluorescent substances by chemical and blowpipe tests.

Several tests for molybdenum depend on whether the mineral occurs as an oxide or as a sulfide.

---

[22]Some prefer to use xanthate crystals with a couple of drops of acid (hydrochloric or sulfuric) instead of the water solution of xanthate.

[23]Instructions for making the bead tests can be found under "Cobalt" or "Chromium."

A sulfide ore must be thoroughly roasted (pulverized and heated on charcoal at a red heat until sulfur fumes are no longer noticeable) before using the mineral in the bead tests.

Tests for molybdenum sulfides

1. Powdered molybdenum sulfide (molybdenite), if heated strongly in an open tube, gives off sulfurous fumes and deposits a pale yellow sublimate (coating) on the sides of the tube, and delicate, hairlike, white or yellow crystals on the mineral itself.

2. Powdered molybdenum sulfide (molybdenite), if heated on charcoal for a long time in the oxidizing (bluish) flame of the blowpipe, deposits a sublimate on the charcoal a short distance from the assay. This sublimate is pale yellow when hot and almost white when cold and often consists of delicate crystals. If this sublimate is touched for an instant with the moderately hot reducing (yellow) flame of the blowpipe, it assumes a beautiful, deep-blue color.

3. Molybdenite is soluble (dissolves) in nitric accid. If heated strongly to dryness in a porcelain receptacle, a beautiful blue coating forms on the bottom and sides of the receptacle.

Tests for molybdenum oxides

1. Finely powdered molybdenum oxides are partially soluble in hydrochloric acid. Tin added to this solution produces a green, blue, and finally a brown-colored solution.

To make this test: Place in a test tube or some other glass or porcelain receptacle a little of the mineral to be tested. Use about the amount of powdered mineral that can be held on the tip of a knife blade. Pour into the receptacle about 2 teaspoonfuls of concentrated (strong) hydrochloric (muriatic) acid, heat to boiling and boil for a few minutes. Dilute this solution by adding about 4 teaspoonfuls of cold water. If some metallic tin is then added to this solution, the solution will turn green, then blue,[24] and finally brown providing the mineral used in the test contained an appreciable amount of molybdenum oxide.

Beginners practicing this test should use wulfenite or some other oxidized molybdenum mineral.

2. Molybdenum oxides can usually be detected by their reduction to the blue oxide of molybdenum.

a) To make this test: Place in a small porcelain crucible or some other porcelain receptacle a little of the powdered mineral to be tested. Use about the amount of powdered mineral that can be held on the tip of a knife blade. Pour onto this mineral a few drops of concentrated sulfuric acid (oil of vitriol), and heat strongly until dense, white fumes of sulfuric acid are given off for a minute or two. Cool the receptacle by blowing strongly onto the mineral. On cooling, a beautiful blue color develops around the sides and in the bottom of the receptacle providing the mineral used in the test contains an appreciable amount of molybdenum oxide.

"The blue color may be of such short duration that the solution seems to turn to brown without showing any blue coloration.

Beginners practicing this test should use wulfenite or some other oxidized molybdenum mineral.

b) This test can also be made in the following manner: Place in a test tube a little of the powdered mineral to be tested. Use about the amount of the powdered mineral that can be held on the tip of a knife blade. Drop onto the mineral a small piece of paper. Add from three to five drops of concentrated sulfuric acid and about an equal amount of water. Heat strongly until dense, white fumes of sulfuric acid are given off freely, and then cool. The liquid on cooling turns a beautiful deep blue color that disappears if the liquid is again heated to boiling, and reappears on cooling providing the mineral used in the test contains an appreciable amount of molybdenum oxide.

Beginners practicing this test should use wulfenite.

NICKEL

Important nickel minerals are millerite (nickel pyrites, sulfuret of nickel, capillary pyrites or nickel sulfide) and niccolite (copper nickel, arsenical nickel or nickel arsenide). Nearly all of the nickel of commerce is obtained from nickel-bearing pyrrohtite and garnierite.

1. In an alkaline solution, dimethylglyoxime[25] solution throws nickel down as a red precipitate.

To make this test: Place in a test tube or some other glass or porcelain receptacle a little of the mineral to be tested. Use about the amount of powdered mineral that can be held on the tip of a knife blade. Pour into the receptacle about 1 teaspoonful of concentrated (strong) nitric acid. Heat until the nickel has been dissolved and then add about 2 teaspoonfuls of cold water to the solution. To this dilute solution add an excess of ammonia (until the solution smells of ammonia). Filter[26] the solution through a good filter and catch the filtrate (the clear solution through the filter paper) in a glass receptacle. To this clear filtrate add a few drops of dimethylglyoxime solution. On the addition of the dimethylgyoxime solution to this filtrate there will be thrown down a light red precipitate providing the mineral used in the test contains an appreciable amount of nickel.

2. Ammonia added to an acid solution of nickel produces a pale blue coloration.

To make this test: Place in a test tube or some other glass or porcelain receptacle a little of the powdered mineral that is to be tested. Use about the amount of powdered mineral that can be held on the tip of a knife blade. Pour into the receptacle about 1 teaspoonful of concentrated nitric acid or a mixture of about 1 teaspoonful of nitric, 1 teaspoonful of hydrochloric (muriatic) acid, and 1 teaspoonful of water. Heat this mixture to boiling

---

[25]Dimethylglyoxime solution is made by dissolving one part of the powder in about ten parts of alcohol. It takes several hours to prepare the solution since the powder dissolves very slowly.

[26]The filtrate must be clear (free from all precipitate).

until the mineral has been dissolved. If the mineral used in the test contains an appreciable amount of nickel, the solution will turn a greenish color, and if an excess of ammonia is added to the solution the solution will turn to a pale blue color that is considerably lighter than that produced by copper.

Beginners practicing this test should use millerite, niccolite, or a piece of metallic nickel.

3. Nickel minerals fused with soda in the reducing flame yield a magnetic button.

To make this test: Mix thoroughly a little of the finely powdered mineral with about twice its volume of sodium carbonate (baking soda). Transfer to a stick of charcoal as much of this mixture as can be held on the tip of a knife blade. Heat strongly before the blowpipe in the reducing (yellow) flame until it is thoroughly fused (melted). The resulting fused mass will contain a dark-colored, more or less metallic button[27] which is magnetic when cold providing the mineral used in this test contains an appreciable amount of nickel.

Beginners practicing this test should use niccolite or millerite.

4. Nickel can usually be detected by the colors it imparts to the fluxes.

a) In the oxidizing (bluish) flame of the blowpipe, borax beads[28] of nickel are violet when warm and brown when cold. In the reducing flame the borax beads are colorless unless saturated with nickel, then they are gray and opaque.

b) The salt of phosphorus beads made in the oxidizing flame are yellow when cold and reddish when warm.

NITRATES

Important natural nitrate minerals are sodium nitrate (soda niter or Chile saltpeter), potassium nitrate (niter or saltpeter), and nitrocalcite (calcium nitrate).

1. Nitrates[29] can usually be detected by the dark blue precipitate in diphenylamine[30] solution.

To make this test: Place in a test tube about 1 teaspoonful of the material to be tested for nitrates. To dissolve the nitrates pour onto this material about 3 tablespoonfuls of water. Heat to boiling, then cool and allow the residue to settle.

Cobalt and iron buttons produced in this manner are also magnetic. For that reason the magnetic button should be further tested for nickel by (1), (2), and (4) and for cobalt and iron by the tests given for those metals.

Instructions for making the bead tests can be found under "Cobalt" and "Chromium."

A sulfide or arsenide ore must be thoroughly roasted (pulverized and heated on charcoal at a red heat, until sulfur or arsenic fumes are no longer noticeable) before using the mineral in the bead tests.

A nitrate is a compound formed by the union of nitric acid with a base, as sodium, potassium, or calcium.

Diphenylamine solution is made by dissolving ½ gram of diphenylamine in 100 cubic centimeters of concentrated sulfuric acid (oil of vitriol) and cautiously adding this to 25 cubic centimeters of water.

Pour into a second test tube about 1 inch of diphenylamine solution. Inclining this test tube slightly, carefully pour about 1 teaspoonful of the liquid from the first receptacle down the inside of the test tube containing the diphenylamine solution. Nitrates if present in the material used in the test usually give a dark blue precipitate with the diphenylamine solution.

Beginners practicing this test should use sodium nitrate, potassium nitrate, or nitric acid.

2. Nitrates can usually be detected by the brown ring formed when a concentrated solution of ferrous sulfate is added to a solution of a nitrate in concentrated sulfuric acid.

To make this test: Place in a test tube or some other glass receptacle a little of the mineral to be tested. Use about ½ teaspoonful of the powdered mineral. Pour into the receptacle about 1 teaspoonful of sulfuric acid (oil of vitriol). Heat this solution to boiling and then cool. If fresh, concentrated ferrous sulfate solution [31] is slowly added to this acid solution a brown ring will form where the ferrous sulfate solution touches the acid solution providing the material used in the test contains an appreciable amount of nitrates. Pungent, brownish-red nitrous oxide fumes are usually given off.

3. Nitrates, when fused with potassium bisulfate, yield brownish-red nitrous oxide fumes.

To make this test: Mix thoroughly a little of the finely powdered mineral with an equal volume of powdered potassium bisulfate (acid sulfate of potassium). Place in a closed tube or test tube about ¼ inch of this mixture. Heat the lower end of the tube at a red heat for some time. Nitrates when given this treatment give off nitrous oxide fumes which are recognized by their reddish-brown color and pungent odor.

Beginners practicing these tests should use nitric acid or sodium nitrate (soda niter or Chile saltpeter).

4. Sodium nitrate (soda niter or Chile saltpeter), potassium nitrate (niter or saltpeter), and calcium nitrate are the nitrate minerals of most commercial interest. All of them are soluble in water and have a salty taste. Sodium nitrate and potassium nitrate give a cooling sensation to the tongue, while potassium nitrate is rather sharp. Nitrocalcite (hydrous calcium nitrate) has a sharp and bitter taste.

OIL SHALES DEFINED

Shales are fine, dense, more or less consolidated sediments that were originally composed principally of fine silt. Their color is usually light gray to black, but yellow, brown, and reddish shales are found in some places. Shales have a noticeably clay-like odor when they are moistened by breathing upon them through the mouth. They can be readily scratched with a knife. The color of powdered shales is much lighter than that of

[31]Concentrated ferrous sulfate solution is made by dissolving that salt in water. Enough of the salt is used so that some of it remains undissolved by the water. If the ferrous sulfate solution is not carefully added, the entire solution may be colored brown.

uncrushed material, and is often nearly white. Shales commonly have a noticeably laminated structure, that is, they appear to be made up of thin sheets or plates. Frequently surface exposures look like piles of cardboard or paper, and such masses can be readily separated into flakes or sheets. Shales are usually associated with sandstones and limestones.

Oil shales have the characteristics above mentioned, but are usually dark brownish-gray to black on freshly broken surfaces and various shades of brown where they have been exposed to the weather. They usually lack all feeling of grittiness between the teeth. Oil shales contain little or no oil as mined, but they do contain variable portions of the solid gum called KEROGEN. Contrary to general belief, oil shales rarely emit the odor of petroleum except sometimes when freshly broken, and even then the odor is faint. Oil shales may be of the decidedly laminated "paper shale" variety, or the material may be a massive, more or less hardened, and very tough clay. Even the latter variety is usually composed of layers of different tints, or in some other way shows its relationship to laminated shales.[30]

This Kerogen is the carbonaceous matter in oil shales from which petroleum (crude oil) is obtained by destructive distillation processes. It can usually be detected if a splinter of the shale is held over a match, candle, or other flame. If the proportion of Kerogen in the shale is fairly high, the splinter will burn for several seconds after it is removed from the flame. As soon as the flame from the Kerogen goes out, white fumes having the odor of burning petroleum are given off for a short time.

ORGANIC MATERIAL

Laboratory studies have brought out important facts regarding sand and other materials that are used for concrete. One of these discoveries is the great importance of being sure that the material is clean, not only in appearance but in fact. Very often, sand which appears to the eye to be clean contains enough humus or vegetable matter to reduce very considerably the strength of the concrete made from it. Tests carried on at Lewis Institute gave the following results:

Concrete made from a clean sand gave a compressive strength at twenty-eight days of 1,900 pounds. Concrete made from this same sand, but with one-tenth of one per cent of tannic acid added, gave a strength of only 1,400 pounds; in other words, one-thousandth part of organic impurities, in terms of the weight of the sand, reduced the strength of the resulting concrete 25 per cent.[31]

We can detect these organic impurities (humus or vegetable matter), even if we cannot see them by ordinary inspection, by using the following colorimetric test for organic impurities, which was devised at the laboratory of the Lewis Institute, Chicago.

To make this test: Take a 12 ounce, graduated prescription bottle and fill to the 4½ ounce mark with the sand to be tested. Pour onto this sand a 3 per cent solution of sodium hydroxide

[30] G. M. Butler and J. B. Tenney, *Petroleum* (Univ. of Ariz., Ariz. Bur. Mines Bull. 130, 1931).
[31] From Lieut. Col. H. C. Boyden's "Notes on Recent Developments in Concrete." For full information on concrete, address Prof. D. A. Abrams of Lewis Institute, Chicago, or The Portland Cement Association, Chicago.

(made by dissolving 1 ounce of sodium hydroxide [caustic soda] in enough water to make 32 ounces [1 quart]), until the volume of the sand and solution, after shaking, amounts to 7 ounces. Shake thoroughly and let it stand for 24 hours. Observe the color of the clear liquid above the sand. If the solution resulting from this treatment is colorless or has a light yellowish color, the sand may be considered satisfactory insofar as organic impurities are concerned. If the liquid is a brown color, especially dark brown, reject the sand or wash it thoroughly before using it for concrete.

PETROLEUM

1. Petroleum (crude oil) in rocks can usually be detected by the odor of petroleum given off when the rocks are vigorously scratched or struck.

2. Petroleum when heated at a high temperature gives off vapors with the characteristic petroleum odor.

To make this test: Fill a test tube or some other small-necked receptacle to about 1 inch from the bottom with the material to be tested for petroleum. If the material is rock, it should be crushed and the pieces used in making this test should pass through a ¼ inch screen. Heat strongly over a flame until gases and vapors are given off freely. Smell the vapors given off. If the material used in the test contains an appreciable amount of petroleum, the gases and vapors given off will have the characteristic odor of petroleum, and petroleum will deposit on the cool portions of the receptacle.

Beginners practicing this test should use a few drops of petroleum.

3. Petroleum can usually be detected by the use of sulfuric ether or chloroform.

To make this test: Fill a test tube or glass bottle to about 1 inch from the bottom with the material to be tested. If the material is rock, it should be crushed, and the pieces used in making the test should pass through a ¼ inch screen. Pour onto this material about 1 inch of sulfuric ether[34] or chloroform. Cork the receptacle tightly. Shake this mixture at intervals until any oil present has been dissolved. (This may take from 1 to 10 hours.) After the petroleum in the material has been dissolved by the sulfuric ether or chloroform, let the mixture stand until the liquid becomes clear. Then pour the clear liquid into a shallow, clean, white china or porcelain dish. The sulfuric ether or chloroform will evaporate quickly leaving a greenish-yellow or brownish ring around the edge of the dish providing the material used in the test contains an appreciable amount of petroleum. The ring, if

---

[34]Sulfuric ether vapors *explode* when ignited. It is *very dangerous* to make this test in the presence of open lights, open flames, or fires. These vapors will anaesthetize (render insensible, put to sleep). The evaporation of these liquids should, therefore, *always* be done in the open, or where the fumes will escape without doing harm.

petroleum, will have an oily feel and the characteristic odor of crude oil.

Beginners practicing this test should use a few drops of petroleum.

PHOSPHORUS

1. Phosphorus can usually be detected by the yellow-colored precipitate formed when a nitric acid solution of phosphorus is added to ammonium molybdate solution.

To make this test: Place in a test tube or some other glass or porcelain receptacle a little of the mineral to be tested. Use about the amount of the powdered mineral that can be held on the tip of a knife blade. Pour into the receptacle about 1 teaspoonful of concentrated (strong) nitric acid and about the same volume of water. Heat this mixture and then cool it. Into another test tube pour about 10 teaspoonfuls of ammonium molybdate solution, and then pour into it a few drops of the cool solution made from the mineral. Let this stand for a time.[35] On the addition of the liquid to the ammonium molybdate solution, a yellow precipitate will be thrown down providing the material used in the test contains an appreciable amount of phosphorus.

Beginners practicing this test should use apatite (phosphate rock or asparagus stone), guano, phosphorite, or some other easily soluble phosphate.[36]

POTASH DEFINED

Potash, properly speaking, is potassium oxide (K_2O), but potassium carbonate (K_2CO_3) is also sometimes (although incorrectly) called "potash." The "potash" salts of commerce do not necessarily contain potassium oxide—for example, potassium chloride (KCl)—and they should more properly be called potassium salts. The term "potash salts," however, is now generally accepted, and all the compounds of potassium are known commercially as potash salts.[37]

From this definition it is evident that the characteristic element of all potash salts is potassium.

POTASSIUM

1. Volatile compounds of potassium color a nonluminous flame violet if heated therein.

To make this test: Wet one end of a piece of platinum or iron wire (baling wire) about 4 inches long with hydrochloric (muriatic) acid so that some of the finely powdered mineral will adhere to it. Draw the wet end of the wire through the powdered mineral. Heat the end of the wire with the mineral on it in a flame. An alcohol lamp flame is very satisfactory for this purpose. As soon as the wire and mineral are red hot, the flame will

---

[35] Sometimes the cold solution must stand for several hours before the precipitate begins to appear.
[36] A phosphate is a salt formed by the combination of phosphoric acid and a base. The determining element of phosphoric acid is phosphorus.
[37] R. B. Ladoo, work cited, p. 437.

be colored violet if the mineral used in the test contains an appreciable amount of volatile compounds of potassium and provided, further, that the potassium flame is not masked[38] or obscured by the flame of some other element.

Beginners practicing this test should use kainite, carnallite, or sylvite.

3. With a Merwin's Flame-Color Screen: Follow the directions outlined in (1), but look at the flame through the different sections of a Merwin's Flame-Color Screen. Through section 1, potassium gives a blue-violet flame, but it appears violet grading into reddish through section 3 and the same tints, but fainter, are seen through section 2.

SILVER

Important silver minerals are native silver, argentite (silver glance, sulfuret of silver, vitreous silver, or silver sulfide), pyrargyite (dark ruby silver, dark red silver, or silver sulfantimonite), proustite (light ruby silver, light red silver, or silver sulfarsenite), and cerargyite (horn silver or silver chloride).

1. Silver can usually be detected by its reduction to metallic silver.

To make this test: Mix thoroughly a little of the finely powdered mineral with about three times its volume of sodium carbonate (baking soda). Transfer to a stick of charcoal as much of this mixture as can be held on the tip of a knife blade. Heat strongly before the blowpipe in the reducing (yellow) flame until it is thoroughly fused (melted). A metallic, silver globule or button results providing the material used in the test contains an appreciable amount of silver. This button is bright when hot or cold and is malleable (it can be flattened out if hammered on an anvil), but it is both harder and less easily cut than a lead button. It should be further tested for silver as explained in the following test "Silver" (2).

Beginners practicing this test should use argentite, pyrargyrite, some other high-grade silver mineral, or metallic silver.

2. Hydrochloric acid and soluble chlorides, when added to a nitric acid solution of silver, give a white precipitate (silver chloride). Silver chloride turns dark on exposure to light and is soluble in ammonia.

To make this test: Place in a test tube or some other glass or porcelain receptacle a little of the mineral to be tested for silver. Use about the amount of the powdered mineral than can be held on the tip of a knife blade. Pour into the receptacle about 2 teaspoonfuls of dilute nitric acid[39] (1 teaspoonful of concentrated

---

[38] The yellow flame of sodium and the red flame of lithium obscure the violet flame of potassium. To intercept and cut off these flames of sodium and lithium, look at the flame through a thick, blue glass.

[39] In test "Silver" (2) it is sometimes better to use concentrated (strong) nitric acid (add no water) for dissolving the silver.

Some silver minerals are insoluble in nitric acid. These minerals if treated as described in "Silver" (1) become easily soluble in that acid.

[strong] nitric acid and about 1 teaspoonful of distilled or rain water). Heat this nitric acid solution to boiling and boil until any silver present has been dissolved. Then cool the solution to room temperature. Add to this cold dilute nitric acid solution a few drops of hydrochloric (muriatic) acid, a small pinch of common table salt, or a few drops of concentrated (strong) salt water. Upon the addition of the hydrochloric acid, the salt, or the salt water to the cold dilute nitric acid solution, a white precipitate[10] will be thrown down providing the material used in the test contains an appreciable amount of silver. If much silver is present, this white precipitate appears as a white curdy mass; if only a small amount of silver is present the precipitate gives the solution a milky appearance. This white precipitate should be tested further for silver by the following tests:

a) Expose some of the solution containing the white precipitate to the bright light for a time. If the white precipitate is silver chloride it will turn dark (violet to brown).

b) Add to the solution containing the white precipitate an excess of ammonia (until the solution smells strong of ammonia). If the white precipitate is silver chloride, the ammonia will dissolve it.

c) If the solution resulting from (b) is rendered acid with nitric acid, the silver will be reprecipitated.

STRONTIUM

The most important strontium minerals are celestite (strontium sulfate) and strontianite (strontium carbonate).

1. Volatile compounds of strontium color a flame crimson if heated therein.

a) To make this test: Use a piece of iron wire (baling wire) about 4 inches long. Wet one end of the wire in dilute (one part acid and four parts of water) hydrochloric (muriatic) acid. Draw the wet end of the wire through the finely powdered mineral. Heat the end of the wire with the mineral on it in a flame. An alcohol lamp flame is very satisfactory for this purpose. As soon as the wire and mineral are red hot, the flame will be colored crimson[11] providing the mineral used in the test contains an appreciable amount of a volatile compound of strontium.

b) Moisten one end of a fragment or chunk of the mineral in hydrochloric acid. Heat[12] the moistened end to red heat at the base of an alcohol lamp flame. As soon as the end of the mineral

---

[10]When these reagents are used in test "Silver" (2), lead and mercury may also be thrown down as a white precipitate. Silver in this form is very soluble in ammonia, while the other two are practically insoluble. It also turns dark on exposure to light. See note under "Lead" (2).

[11]The crimson flame of strontium must not be mistaken for the red flame of lithium. The flame may be of such short duration that it appears as a crimson flash.

[12]If the fragment of mineral used in the test is small, use a clean pair of iron tweezers, pliers, or pincers for holding it in the flame.

is ignited (red hot) the flame will be colored crimson providing the mineral used in the test contains an appreciable amount of a volatile compound of strontium.

2. With Merwin's Flame-Color Screen: Follow the directions outlined in "Strontium" (1) but look at the flame through the different sections of a Merwin's Flame-Color Screen. Through section 3, strontium gives a crimson-colored flame which is absorbed by sections 1 and 2.

3. Some strontianites (strontium carbonate) fluoresce (glow) a white or green color when exposed to strong ultraviolet rays (black light)—best seen in the dark.

Always check fluorescent substances by chemical and blowpipe tests.

TIN

The only important tin mineral is cassiterite (tinstone or tin dioxide; when it is recovered from placers it is called stream tin). Cassiterite (tin dioxide) when boiled with metalliz zinc in hydrochloric or sulfuric acid usually becomes coated with a gray metallic deposit.

To make this test: Place in a test tube a ragged fragment of the mineral, somewhat larger than a bean. Around this fragment pour metallic zinc, allowing about one half of the fragment to project above the zinc. Granulated zinc, about 20 mesh, works satisfactorily in this test. Onto this pour about 3 teaspoonfuls of either concentrated (strong) hydrochloric (muriatic) acid or concentrated sulfuric (oil of vitriol) acid and about 2 teaspoonfuls of water. Heat this mixture to boiling and allow it to boil for a couple of minutes.

Cassiterite when given this treatment usually becomes coated with a dull gray metallic deposit. If the fragment is washed in water and the coating is then rubbed dry the metallic deposit becomes bright.

2. Tin can usually be detected by the sublimate formed on charcoal and its reduction to small, metallic globules. These globules, if treated with nitric acid, yield a white powder.

To make this test: Mix thoroughly a little of the finely powdered mineral with an equal volume of powdered charcoal and two volumes of sodium carbonate (baking soda). Transfer to a shallow cavity in a stick of charcoal about the amount of this mixture than can be held on the tip of a knife blade, and form a paste of the mixture with water. Heat before the blowpipe in a strong, reducing (yellow) flame. This treatment gives a scanty sublimate (coating) on the charcoal which is yellowish when hot and white when cold.

If this sublimate is moistened with a drop or two of cobalt nitrate solution and if the assay is then heated strongly before the blowpipe in the reducing flame, the sublimate will assume a dull, bluish-green color when cold.

This treatment also yields small metallic globules, which can only with difficulty be forced to run together into one single, larger globule. These globules on cooling become coated with a white film but if cut open show a white, metallic color. If these globules are treated with nitric acid, a white powder is produced which is insoluble in that acid.

TUNGSTEN

Important tungsten minerals are wolframite[13] (wolfram, hübnerite, megabasite, or tungstate of iron and manganese), ferberite (iron tungstate), tungstite (sulfide), scheelite (calcium tungstate), and cuproscheelite (calcium copper tungstate).

1. In hydrochloric acid, tungsten gives a lemon-yellow residue that is soluble in ammonia.

To make this test: Place in a test tube or some other glass or porcelain receptacle a little of the finely powdered mineral to be tested. Use about the amount of the powdered mineral that can be held on the tip of a knife blade. Pour into the receptacle about 2 teaspoonfuls of concentrated (strong) hydrochloric (muriatic) acid. Heat this mixture to boiling and boil for a time, replenishing the acid if necessary. This treatment will give a lemon-yellow residue (tungstic acid) in the bottom of the receptacle providing the mineral used in the test contains an appreciable amount of tungsten. This lemon-yellow residue, if it is tungstic acid, is soluble in ammonia. Therefore this test should be carried further for tungsten by testing the solubility of the yellow residue in ammonia. To do so, pour into the receptacle an excess of ammonia (until the solution smells strong of ammonia). Warm slightly if necessary. The lemon-yellow residue will dissolve in the ammonia if it is tungstic acid (tungsten).

Beginners practicing this test should use scheelite or some other easily soluble tungsten mineral.

2. In hydrochloric acid solutions of tungsten, tin or zinc give a blue color.

To make this test: Pour into a test tube or some other glass or porcelain receptacle a little of the finely powdered mineral to be tested. Use about the amount of the mineral that can be held on the tip of a knife blade. Pour into the receptacle about 2 teaspoonfuls of concentrated hydrochloric acid. Heat this mixture to boiling and boil strongly for a time, replenishing the acid if necessary. After a lemon-yellow colored residue begins to form in the bottom of the receptacle, add a little metallic tin or metallic zinc to the solution. If no other tin or zinc is available, use a piece of galvanized iron with zinc on it, a piece of a tin can with tin on it, soft solder with tin in it, or hard solder with zinc in it. On addition of the tin or zinc to this solution, the solution will turn

---

[13]Tests "Tungsten" (1) and (2) should be used only on very soluble tungsten minerals, as these tests take too long if used directly on rather insoluble minerals. The rather insoluble minerals should be given the preliminary fusing treatment as described in "Iron" (2). The fused mass resulting from the fusion should then be tested as described in tests "Tungsten" (1) and (2).

deep blue and, later, brown providing the material used in the test contains an appreciable amount of tungsten.

Beginners practicing this test should use scheelite or some other easily soluble tungsten mineral.

3. Tungsten can usually be detected by the colors it imparts to the fluxes.

a) All of the borax beads[11] of tungsten are colorless except the warm, saturated beads which are yellow.

b) All of the salt of phosphorus beads of tungsten are colorless except the warm, saturated bead which is yellow and the cold, saturated bead, made in the reducing flame, which is greenish blue.

4. Some scheelites (calcium tungstate) fluoresce (glow) a blue or bluish-white color when exposed to strong ultraviolet rays (black light).

Always check fluorescent substances by chemical and blowpipe tests.

VANADIUM

Important vanadium minerals are patronite (vanadium sulfide), vanadiferous asphaltite, vanadinite (chlorovanadate of lead) descloizite (hydrated basic vanadate of lead and zinc), roscoelite (vanadium mica), and carnotite (a mixture of vanadium and uranium compounds).

1. Vanadium colors concentrated hydrochloric acid a deep cherry red. This reaction takes place with the evolution of chlorine gas. A little water added to this red solution changes it to a light green color.

To make this test: Place in a bone-dry test tube or some other glass or porcelain receptacle a little of the mineral to be tested. Use about the amount of the powdered mineral that can be held on the tip of a knife. Pour onto the mineral about ½ teaspoonful of concentrated (strong) hydrochloric (muriatic) acid. Almost as soon as the hydrochloric acid comes in contact with the mineral, chlorine gas will be given off, and the solution will turn to a deep, cherry-red color providing the mineral used in the test contains an appreciable amount of vanadium. This chlorine gas is easily recognized by its rusty-green color and its pungent odor. If a few drops of water are added to this cherry-red solution, it changes to a light greenish tint. If too much water is added, the solution becomes almost colorless.

Beginners practicing this test should use vanadinite, descloizite, or some other vanadate.

2. Vanadium can usually be detected by the greenish color given sulfuric acid.

To make this test: Place in a test tube or some other glass or porcelain receptacle a little of the mineral that is to be tested. Use about the amount of powdered mineral that can be held on

---

[11]Instructions for making the bead tests can be found under "Cobalt" and "Chromium."

the tip of a knife blade. Pour onto the mineral about 1 teaspoonful of concentrated sulfuric acid (oil of vitriol). Heat the
solution to boiling and boil until dense, white fumes of sulfuric
acid are given off. Cool to room temperature and when cold add
very carefully from 1 to 2 teaspoonfuls of cold water. Almost
immediately upon the addition of the water the color of the solution will change to a light green providing the mineral used in the
test contains an appreciable amount of soluble vanadium.

Beginners practicing this test should use vanadinite, descloizite, or some other vanadate.

3. Vanadates with potassium bisulfate in a closed tube give a
yellow mass.

To make this test: Mix thoroughly a little of the finely powdered mineral with an equal volume of potassium bisulfate (acid
sulfate of potassium). Place in a closed tube about ½ inch of this
mixture. Heat the lower end of the tube at a red heat for some
time. This will produce a yellow mass providing the mineral
used in the test contains vanadates.

Beginners practicing this test should use vanadinite, descloizite, or some other vanadate.

4. Hydrogen peroxide gives a brownish-red color to an acid
solution of vanadium.

To make this test: Place in a porcelain receptacle a little of
the mineral to be tested. Use about the amount of the powdered
mineral that can be held on the tip of a knife blade. Pour into
the receptacle about 1 teaspoonful of concentrated hydrochloric
acid and boil the solution for a couple of minutes. Cool the receptacle and pour into it about 1 teaspoonful of concentrated
nitric acid. Boil for a couple of minutes and cool. After it is
cold add about 1 teaspoonful of concentrated sulfuric acid and
boil this mixture over an open flame until dense white fumes are
given off. Cool to room temperature and very carefully add about
3 teaspoonfuls of cold water. Boil for a couple of minutes and
then filter. Catch the filtrate (the clear solution that passes
through the filter paper) in a glass or porcelain receptacle. Pour
about 1 teaspoonful of this filtrate into a test tube. To this add a
drop or two of fresh hydrogen peroxide. On the addition of the
hydrogen peroxide, the solution will become brownish red providing the mineral used in the test contained an appreciable
amount of soluble vanadium.

Beginners practicing this test should use vanadinite or some
other vanadate.

5. Vanadium[45] can usually be detected by the colors it imparts to the fluxes.

a) In the oxidizing (bluish) flame of the blowpipe, borax

---

[45]A sulfide or arsenide ore must be thoroughly roasted (pulverized and
heated on charcoal at a red heat until sulfur or arsenic fumes are no
longer noticeable) before using the mineral in the bead test.

beads[46] of vanadium are all yellow except the unsaturated, cold bead, which is colorless. In the reducing (yellow) flame all borax beads of vanadium are green.

b) All salt of phosphorus beads of vanadium made in the oxidizing flame are yellow, and all such beads made in the reducing flame are green.

6. Some vanadinites (lead vanadate) fluoresce (glow) a green color when exposed to strong ultraviolet rays (black light).

Always check fluorescent substances by chemical and blowpipe tests.

ZINC

Important zinc minerals are sphalerite (blende, zinc blende, blackjack, false lead, false galena, or zinc sulfide), smithsonite (dry-bone ore or zinc carbonate), and calamine (electric calamine, hydrous zinc silicate, basic zinc metasilicate or silicate of zinc).

1. Zinc[47] can usually be detected by the sublimate formed on charcoal when heated with soda before the blowpipe. This sublimate is yellow when hot and white when cold. If moistened with cobalt nitrate solution and heated, it assumes a green color.

To make this test: Mix thoroughly a little of the finely powdered mineral with an equal volume of sodium carbonate (baking soda). Transfer to a shallow cavity in a stick of charcoal about the amount of this mixture that can be held on the tip of a knife blade, and form a paste of the mixture by moistening with water. Heat before the blowpipe in a strong, reducing (yellow) flame. This treatment gives a scanty sublimate (coating) on the charcoal which is canary-yellow colored when hot and white when cold, providing the mineral used in the test contains an appreciable amount of zinc.

If the sublimate made in the test just described is moistened with a drop or two of cobalt nitrate solution and if the assay is then heated strongly before the blowpipe in the reducing flame, the sublimate will assume a bright green color that is best seen when it is cold.

Beginners practicing this test should use sphalerite, smithsonite, or a small piece of metallic zinc.

2. Zinc is thrown down as a white precipitate by ammonium sulfide from an alkaline solution, thus being the only white sulfide that is insoluble in such a solution.

To make this test: Place in a test tube or some other glass or porcelain receptacle a little of the mineral that is to be tested. Use about that amount of the powdered mineral that can be held on the tip of a knife blade. Pour onto the mineral in the receptacle about 1 teaspoonful of concentrated (strong) hydrochloric

---

[46]Instructions for making the bead tests can be found under "Cobalt" and "Chromium."

[47]A few zinc compouds such as sphalerite give the above results without mixing the mineral with a flux.

(muriatic) acid and a drop of nitric acid, and heat to boiling. After the zinc has been dissolved, add about 2 teaspoonfuls of cold water and cool the solution to room temperature. To the cold solution add an excess of ammonia (until the solution smells strong of ammonia). Any iron present will be thrown down as a brownish-red precipitate as mentioned in test "Iron" (3). Filter off the residue and precipitate, and catch the clear, filtered solution in another glass or porcelain receptacle. To the clear, filtered solution add a few drops of ammonium sulfide solution. This will throw down zinc as a white precipitate (zinc sulfide).

3. Silicates of zinc when moistened with cobalt nitrate and heated before the blowpipe assume a blue color. The following method for making this test is given in G. M. Butler's *Handbook of Blowpipe Analysis.*

To make this test:

Hold a small splinter of the substance to be tested in the platinum forceps and heat it in the blowpipe flame to the highest possible temperature. Then examine it with a lens; if it shows any signs of fusion, this test cannot be applied. If non-fusible, moisten it with cobalt nitrate and ignite strongly in the hottest part of the blowpipe flame. It will first turn black, but after a prolonged heating may assume a characteristic tint. If a splinter of the substance cannot be obtained, it should be powdered and the test conducted upon a flat cake of the powder upon charcoal. Longer heating is required by this method, however, and the results are not apt to be as satisfactory.

This test can be applied only to nonfusible, white or faintly tinted minerals, or those which become white or faintly tinted upon ignition.

A blue coloration best seen when cold indicates zinc, but infusible aluminum minerals will yield the same color when treated in this way.

Beginners practicing this test should use calamine or willemite (zinc silicates).

4. Some zinc minerals (some willemites, smithsonites, sphalerites, and others) fluoresce (glow) when exposed to strong ultraviolet rays (black light). Green, yellow, and blue are among the colors emitted by these minerals when activated by this light—best seen in the dark.

Always check fluorescent substances by chemical and blowpipe tests.

APPENDIX I

FEDERAL STATUTES GOVERNING SIZES OF MINING CLAIMS

LODE CLAIMS

Length of lode claim

Not to exceed 1,500 feet.—Revised Statutes Section 2320.—Mining claims upon veins or lodes of quartz or other rock in place bearing gold, silver, cinnabar, lead, tin, copper, or other valuable deposits, heretofore located, shall be governed as to length along the vein or lode by the customs, regulations, and laws in force at the date of their location. A mining claim located after the tenth day of May, 1872, whether located by one or more persons, may equal, but shall not exceed, 1,500 feet in length along the vein or lode; but no location of a mining claim shall be made until the discovery of the vein or lode within the limits of the claim located.—Section 2, Act of Congress, May 10, 1872.

Width of lode claim

Revised Statutes Section 2320.—No claim shall extend more than 300 feet on each side of the middle of the vein at the surface, nor shall any claim be limited by any mining regulation to less than 25 feet on each side of the middle of the vein at the surface except where adverse rights existing on the tenth day of May, 1872, render such limitation necessary. The end-lines of each claim shall be parallel to each other.—Section 2, Act of Congress, May 10, 1872.

The following interpretations are drawn from the above:

Lode claim.—It cannot exceed 1,500 feet in length, along the vein or lode. It cannot extend more than 300 feet nor less than 25 feet on each side of the middle of the vein or lode. The end lines must be parallel to each other. The maximum area of a lode claim is 20.6 acres.

PLACER CLAIMS

Open to location and patent

Revised Statute Section 2329.—Claims usually called "placers," including all forms of deposit, excepting veins of quartz, or other rock in place, shall be subject to entry and patent, under like circumstances and conditions, and upon similar proceedings, as are provided for vein or lode claims; but where the lands have been previously surveyed by the United States, the entry in its exterior limits shall conform to the legal subdivisions of the public lands.—Section 12, Act of Congress, July 9, 1870.

Size of claim

Revised Statutes Section 2330.—Legal subdivisions of 40 acres may be subdivided into 10-acre tracts; and two or more persons, or associations of persons, having contiguous claims of any size, although such claims may be less than 10 acres each, may make joint entry thereof; but no location of a placer-claim made after the ninth day of July, 1870, shall exceed 160 acres for any one person or association of persons, which location shall conform to the United States surveys; and nothing in this section contained shall defeat or impair any bona-fide pre-emption or homestead claim upon agricultural lands, or authorize the sale of the improvements of any bona-fide settler to any purchaser.

Twenty acres to one locator

Revised Statutes Section 2331.—Where placer claims are upon surveyed lands, and conform to legal subdivisions, no further survey or plat shall be required, and all placer-mining claims located after the tenth day of May, 1872, shall conform as near as practicable with the United States system of public-land surveys, and the rectangular subdivisions of such surveys, and no such location shall include more than 20 acres for each individual claimant; but where placer claims can not be conformed to legal subdivisions, survey and plat shall be made as on unsurveyed lands; and where by the segregation of mineral land in any legal subdivision a quantity of agricultural land less than 40 acres remains, such fractional portion of agricultural land may be entered by any party qualified by law, for homestead or pre-emption purposes.—Section 10, Act of Congress, May 10, 1872.

The following interpretations are drawn from the above:

Placer claim.—The maximum size of a placer claim for each claimant is 20 acres. Placer claim boundaries conform to public land surveys.

MILL SITES

Extent—how patented

Revised Statute Section 2337.—Where non-mineral land not contiguous to the vein or lode is used or occupied by the proprietor of such vein or lode for mining or milling purposes, such non-adjacent surface-ground may be embraced and included in an application for a patent for such vein or lode, and the same may be patented therewith, subject to the same preliminary requirements as to survey and notice as are applicable to veins or lodes; but no location hereafter made of such non-adjacent land shall exceed 5 acres, and payment for the same must be made at the same rate as fixed by this chapter for the superficies of the lode. The owner of a quartz-mill or reduction-works, not owning a mine in connection therewith, may also receive a patent for his millsite, as provided in this section.—Section 15, Act of Congress, May 10, 1872.

APPENDIX II

MINING STATUTES
FROM
THE REVISED CODE OF ARIZONA
1928

ARTICLE 1.—MINING LOCATION

2266.—Location upon discovery of mineral in place. On the discovery of mineral in place on the public domain of the United States the same may be located as a mining claim by the discoverer for himself, or for himself and others, or for others.

2267.—Location notice, contents; amendment. Such location shall be made by erecting at or contiguous to the point of discovery a conspicuous monument of stones, not less than 3 feet in height, or an upright post, securely fixed, projecting at least 4 feet above the ground, in or on which there shall be posted a location notice, signed by the name of the locator. The location notice must contain: The name of the claim located; the name of the locator; the date of the location; the length and width of the claim in feet, and the distance in feet from the point of discovery to each end of the claim; the general course of the claim; the locality of the claim with reference to some natural object or permanent monument whereby the claim can be identified; and until each of the same shall have been done no right to such location shall be acquired. The notices may be amended at any time and the monuments changed to correspond with the amended location; provided, that no change shall be made which will interfere with the rights of others.

2268.—Completing location; additional facts; failure. From the time of the location, the locator shall be allowed ninety days within which to do the following: Cause to be recorded in the office of the county recorder a copy of the location notice; sink a discovery shaft in the claim to a depth of at least 8 feet from the lowest part of the rim of the shaft at the surface, and deeper, if necessary, until there is disclosed in said shaft mineral in place; and monument the claim on the ground so that its boundaries can be readily traced. Failure to do all such things in the time and place specified shall be an abandonment of the claim, and all right and claim thereto of the discoverer and locator shall be forfeited.

2269.—Monumenting. Such boundaries shall be monumented by six substantial posts projecting at least 4 feet above the surface of the ground, or by substantial stone monuments at least 3 feet high, one at each corner of the said claim and one at the center of each end line thereof; when, however, the point of a monument is at the same point and coincides with a monument of the survey of the United States, the monument of such government survey shall be deemed a mining claim monument.

2270.—Tunnel as location work. Any open cut, adit or tunnel, made as a part of the location of a lode mining claim, equal in amount of work to a shaft 8 feet deep and 4 feet wide by 6 feet long, and which shall cut a lode or mineral in place at a depth of 10 feet from the surface, shall be equivalent, as discovery work, to a shaft sunk from the surface.

2271.—Relocation. The location of an abandoned or forfeited claim shall be made in the same manner as other locations, except that the relocator may perform his location work by sinking the original location shaft 8 feet deeper than it was originally, or if the original location work consisted of a tunnel or open cut, he may perform his location work by extending said tunnel or open cut by removing therefrom 240 cubic feet of rock or vein material.

2272.—Locating and monumenting placer claims; recording notice. The locator of a placer mining claim shall locate his claim in the following manner: By posting a location notice thereon containing the name of the claim, the name of the locator, the date of location and the number of acres claimed; a description of the claim with reference to some natural object or permanent monument that will identify the claim and by marking the boundaries of his claim with a post or monument of stones at each angle of the claim located. When a post is used it must be at least 4 inches in diameter by 4 feet 6 inches in length, set one foot in the ground and surrounded by a mound of stone or earth. Where it is practically impossible on account of a bed of rock or precipitous ground to sink such posts, they may be placed in a pile of stones. If it is impossible to erect and maintain a post or monument of stone at any angle of such claim, a witness post or monument may be used, to be placed as near the true corner as the nature of the ground will permit. When a mound of stone is used, it must be at least 3 feet in height and 4 feet in diameter at the base. The locator shall, within sixty days after the date of location, record a copy of the location notice.

2273.—Affidavit of performance of annual work; prima facie evidence. Within three months after the expiration of the time fixed for the performance of annual labor or the making of improvements upon a mining claim, the person on whose behalf such work or improvement was made, or some person for him knowing the facts, may make and record an affidavit, in substance, as follows:

State of Arizona, county of....................ss: ....being duly sworn, deposes and says that he is a citizen of the United States and more than twenty-one years of age, resides at.............., in.. ....county, Arizona, is personally acquainted with the mining claim known as...................... mining claim, situated in ....mining district, Arizona, the location notice of which is recorded in the office of the county recorder of said county, in book...........of records of mines, at page............... That between the...........day of..........., A.D..........., and the...........day of ...., A.D..........., at least...........dollars worth of work and improvements were done and performed upon said claim, not including the location work of said claim. Such work and improvements were made by and at the expense of........................, owners of said claim, for the purpose of complying with the laws of the United States pertaining to assessment or annual work, and (here name the miners or men who worked upon the claim in doing the work) were the men employed by said owner and who labored upon said claim, did said work and improvements, the same being as follows, to wit: (Here describe the work done, and add signature and verification).

Such affidavit when recorded, shall be prima facie evidence of the performance of such labor or improvements.

2274.—One affidavit for group. When two or more contiguous claims are owned by the same person, and constitute a group, and the annual work is done upon each of said claims or upon one of more of the same for the benefit of all, or wholly or partly outside of such claims for the benefit of all, all such claims may be included in a single affidavit.

2277.—Sufficiency of description of mining claims. In all actions, judgments, grants or conveyances it shall be sufficient description of a mining claim if the name of the claim, the district, county and state where it is situate, and the book and page where the location notice thereof is recorded can be intelligently learned therefrom.

APPENDIX III

TABLES OF WEIGHTS AND MEASURES
WEIGHTS

AVOIRDUPOIS WEIGHT

1 Dram = 27.343 Grains.
1 Ounce = 16 Drams = 437.5 Grains.
1 Pound = 16 Ounces = 7,000 Grains.
1 Short Ton = 2,000 Pounds = 32,000 Ounces = 0.8928 Long Ton.
1 Long Ton = 2,240 Pounds = 35,840 Ounces = 1.12 Short Tons.

TROY WEIGHT*

(Used in Weighing Gold or Silver)

1 Pennyweight = 24 Grains.
1 Ounce = 20 Pennyweights = 480 Grains.
1 Pound = 12 Ounces = 5,760 Grains.

APOTHECARIES' WEIGHT*

1 Scruple = 20 Grains.
1 Dram = 3 Scruples = 60 Grains.
1 Ounce = 8 Drams = 480 Grains.
1 Pound = 12 Ounces = 5,760 Grains.

METRIC MEASURE

1 Centigram = 10 Milligrams.
1 Decigram = 10 Centigrams.
1 Gram = 10 Decigrams = 0.001 Kilogram.
1 Decagram = 10 Grams = 0.01 Kilogram.
1 Hectogram = 10 Decagrams = 100 Grams = 0.1 Kilogram.
1 Kilogram = 10 Hectograms = 1,000 Grams.
1 Metric Ton = 1,000 Kilograms = 1,000,000 Grams.

EQUIVALENTS

1 Ounce (Avoir.) = 0.911 Ounce (Troy) = 28.35 Grams.
1 Ounce (Troy or Apoth.) = 1.097 Ounces (Avoir.) = 31.103 Grams.
1 Pound (Avoir.) = 1.215 Pounds (Troy or Apoth.) = 14.58 Ounces (Troy or Apoth.) = 0.4536 Kilogram = 453.6 Grams.
1 Pound (Troy or Apoth.) = 0.82286 Pound (Avoir.) = 13.166 Ounces (Avoir.) = 0.3732 Kilogram = 373.2 Grams.
1 Short Ton = 0.8928 Long Ton = 0.9072 Metric Ton = 907.2 Kilograms.
1 Long Ton = 1.12 Short Tons = 1.016 Metric Tons = 1,016 Kilograms.
1 Gram = 0.0353 Ounce (Avoir.) = 0.03215 Ounce (Troy or Apoth.) = 15.432 Grains.

---

*The Troy grain, ounce, and pound weigh the same as the respective units of the Apothecaries' system.

1 Kilogram = 2.2046 Pounds (Avoir.) = 2.679 Pounds (Troy or Apoth.) = 35.274 Ounces (Avoir.) = 32.151 Ounces (Troy or Apoth.).

1 Metric Ton = 1.1023 Short Tons = 0.984 Long Ton = 2,204.6 Pounds (Avoir.).

MEASURES OF LENGTH

LINEAR MEASURE

1 Span = 9 Inches.

1 Foot = 12 Inches = 0.3333 Yard = 0.000189 Statute Mile.

1 Yard = 3 Feet = 36 Inches = 0.000568 Statute Mile.

1 Fathom = 2 Yards = 6 Feet = 72 Inches = 0.001136 Statute Mile.

1 Rod = 5.5 Yards = 16.5 Feet = 198 Inches = 0.003125 Statute Mile.

1 Furlong = 40 Rods = 220 Yards = 660 Feet = 7,920 Inches = 0.125 Statute Mile.

1 Statute Mile = 8 Furlongs = 1,760 Yards = 5,280 Feet = 63,360 Inches.

GUNTER'S CHAIN

1 Link = 7.92 Inches.

1 Chain = 100 Links = 4 Rods.

1 Statute Mile = 80 Chains = 320 Rods = 8,000 Links.

METRIC MEASURE

1 Centimeter = 10 Millimeters.

1 Decimeter = 10 Centimeters = 100 Millimeters.

1 Meter = 10 Decimeters = 100 Centimeters.

1 Decameter = 10 Meters = 1,000 Centimeters.

1 Hectometer = 10 Decameters = 100 Meters = 10,000 Centimeters.

1 Kilometer = 10 Hectometers = 1,000 Meters = 100,000 Centimeters.

EQUIVALENTS

1 Inch = 0.0833 Foot = 2.54 Centimeters = 0.0254 Meter.

1 Foot = 30.48 Centimeters = 0.3048 Meter.

1 Yard = 91.44 Centimeters = 0.9144 Meter.

1 Statute Mile = 1.6094 Kilometers = 1,609.4 Meters.

1 Centimeter = 0.3937 Inch = 0.0328 Foot.

1 Meter = 39.37 Inches = 3.28 Feet = 1.0936 Yards = 0.00062 Statute Mile.

1 Kilometer = 0.6214 Statute Mile = 1,093.6 Yards = 3,281 Feet.

MEASURES OF SURFACE

SQUARE OR LAND MEASURE

1 Square Inch = 0.0069 Square Foot.
1 Square Foot = 144 Square Inches = 0.111 Square Yard.
1 Square Yard = 9 Square Feet = 1,296 Square Inches.
1 Square (Architect's Measure) = 100 Square Feet.
1 Square Rod = 30.25 Square Yards = 272.25 Square Feet = 39,204 Square Inches.
1 Acre = 160 Square Rods = 4,840 Square Yards = 43,560 Square Feet = 0.00156 Square Mile.
A surface 208.71 feet long by 208.71 feet wide = 1 Acre.
1 Square Mile (Section) = 640 Acres = 3,097,600 Square Yards = 27,878,400 Square Feet.

METRIC SQUARE MEASURE

1 Square Centimeter = 100 Square Millimeters.
1 Square Decimeter = 100 Square Centimeters.
1 Square Meter (Centiare) = 100 Square Decimeters = 10,000 Square Centimeters.
1 Square Decameter (Are) = 100 Centiares (Square Meters) = 1,000,000 Square Centimeters.
1 Square Hectometer (Hectare) = 100 Ares = 10,000 Centiares (Square Meters).
1 Square Kilometer = 100 Hectares = 10,000 Ares = 1,000,000 Centiares.

EQUIVALENTS

1 Square Inch = 6.45 Square Centimeters.
1 Square Foot = 0.0929 Centiare (Square Meter) = 929 Square Centimeters.
1 Square Yard = 0.8361 Centiare (Square Meter).
1 Acre = 0.4047 Hectare (Square Hectometer) = 4,046.9 Centiares (Square Meters).
1 Square Mile = 2.59 Square Kilometers = 259 Hectares.
1 Square Centimeter = 0.155 Square Inch.
1 Centiare (Square Meter) = 1.196 Square Yards = 10.764 Square Feet = 1,550 Square Inches.
1 Hectare = 2.471 Acres = 0.003861 Square Statute Mile.
1 Square Kilometer = 247.1 Acres = 0.3861 Square Statute Mile.

MEASURES OF VOLUME

U.S.A. CUBIC MEASURE

1 Cubic Inch = 0.000578 Cubic Foot.
1 Cubic Foot = 0.037 Cubic Yard = 1.728 Cubic Inches.
1 Cubic Yard = 27 Cubic Feet = 46.656 Cubic Inches.

METRIC CUBIC MEASURE

1 Cubic Centimeter = 1,000 Cubic Millimeters = 1 Milliliter.
1 Cubic Decimeter (Liter) = 1,000 Cubic Centimeters.
1 Cubic Meter (Stere) = 1,000 Cubic Decimeters (Liters) = 1,000,000 Cubic Centimeters.

EQUIVALENTS

1 Teaspoonful (Liquid) = Approximately 5 Cubic Centimeters.
1 Tablespoonful (Liquid) = Approximately 15 Cubic Centimeters.
1 Cubic Inch = 16.39 Cubic Centimeters.
1 Cubic Foot = 28,320 Cubic Centimeters = 0.028 Cubic Meter.
1 Cubic Yard = 0.764 Cubic Meter.
1 Cubic Centimeter = 0.061 Cubic Inch.
1 Cubic Meter = 1.307 Cubic Yards = 35.31 Cubic Feet = 61,020 Cubic Inches.
1 Cord of Wood = 128 Cubic Feet (cut 4 feet wide, piled 4 feet high and 8 feet long).

U.S.A. DRY MEASURE

1 Pint = 0.5 Quart = 0.125 Gallon = 0.0625 Peck = 0.0156 Bushel.
1 Quart = 2 Pints = 0.25 Gallon = 0.125 Peck = 0.0312 Bushel.
1 Gallon = 4 Quarts = 8 Pints = 0.5 Peck = 0.125 Bushel.
1 Peck = 2 Gallons = 8 Quarts = 16 Pints = 0.25 Bushel.
1 Bushel = 4 Pecks = 8 Gallons = 32 Quarts = 64 Pints.

LIQUID MEASURE

1 Gill = 0.25 Pint.
1 Pint = 4 Gills = 0.5 Quart = 0.125 U.S. Gallon.
1 Quart = 2 Pints = 8 Gills = 0.25 U.S. Gallon = 0.0079 Barrel.
1 U.S. Gallon = 4 Quarts = 8 Pints = 32 Gills = 0.0317 Barrel.
1 Barrel = 31.5 U.S. Gallons = 126 Quarts = 252 Pints = 1,008 Gills.

There is no standard liquid "Barrel." The one used in this table contains 35.31 cubic feet.

APOTHECARIES' FLUID MEASURE

1 Fluid Dram = 60 Minims.
1 Fluid Ounce = 8 Fluid Drams.
1 Fluid Pint = 16 Fluid Ounces = 1 Pint (Liquid Measure).
1 Gallon = 8 Pints = 128 Fluid Ounces.

METRIC LIQUID MEASURE

1 Centiliter = 10 Milliliters.
1 Deciliter = 10 Centiliters = 100 Milliliters.
1 Liter (Cubic Decimeter or Millistere) = 10 Deciliters = 100 Centiliters = 1,000 Milliliters (Cubic Centimeters) = 0.001 Cubic Meter.
1 Decaliter (Centistere) = 10 Liters = 1,000 Centiliters.
1 Hectoliter (Decistere) = 10 Decaliters = 100 Liters.
1 Kiloliter (Stere) = 10 Hectoliters = 1,000 Liters.

EQUIVALENTS

1 Quart (U.S. Dry) = 67.2 Cubic Inches = 0.0389 Cubic Foot
= 0.0011 Cubic Meter = 1.1 Liters = 1,101 Cubic Centimeters.

1 Quart (U.S. Liquid) = 57.75 Cubic Inches = 0.033 Cubic Foot
= 0.00094 Cubic Meter = 0.946 Liter = 946 Cubic Centimeters.

1 Gallon (U.S. Dry) = 268.8 Cubic Inches = 0.1556 Cubic Foot
= 0.004 Cubic Meter = 4.405 Liters = 4,405 Cubic Centimeters.

1 Gallon (U.S. Liquid) = 231 Cubic Inches = 0.1337 Cubic Foot
= 0.00378 Cubic Meter = 3.785 Liters = 3,785 Cubic Centimeters.

1 Liter = 1.0567 (U.S. Liquid) Quarts = 0.908 (U.S. Dry) Quart
= 0.0353 Cubic Foot = 61.023 Cubic Inches = 33.81 Apoth.
Fluid Ounces = 0.264 Gallon (U.S. Liquid) = 0.227 Gallon
(U.S. Dry).

AVERAGE WEIGHTS OF VARIOUS SUBSTANCES

Name	Pounds per cubic foot	Cubic feet per short ton (2,000 lbs.)
Aluminum, cast	160	12.5
Amalgam	868	2.3
Andesite	181	11.0
Antimony, cast	418	4.8
Antimony sulfide (stibnite)	287	7.0
Arsenic	357	5.6
Arsenic sulfide (realgar)	218	9.2
Asbestos	175	11.4
Asphaltum	69 to 75	26.5 to 29.0
Barium	242	8.2
Barium sulfate (barite)	280	7.1
Basalt (traprock)	181	11.0
Borax	109	18.3
Brass (copper and zinc), cast	527	3.7
Brick, common	100 to 130	15.4 to 20.0
Bronze (aluminum)	480	4.1
Cadmium	539	3.7
Calcite	168	11.8
Calcium	99	21.0
Cement (Portland)	85 to 195	10.2 to 23.6
Chalk	146	13.7
Chromium	312	6.4
Clay, loose, dry	65	30.8
Coal, bituminous, broken loose	47 to 60	36.4 to 42.5
Cobalt, nickel arsenide (smaltite)	405	4.9
Concrete, stone	130 to 150	13.3 to 15.4
Copper, cast	550	3.6
Copper carbonate (malachite)	243	8.2
Copper pyrites (chalcopyrite)	262	7.6
Diabase or diorite	187	10.6

BIBLIOGRAPHY

Boyden, H. C., *Notes on Recent Developments in Concrete*, The Portland Cement Association, Chicago, Ill.

Brush, G. J., and Penfield, S. L., *Determinative Mineralogy*, John Wiley & Sons, Inc., New York.

Butler, G. M., *Handbook of Blowpipe Analysis*, John Wiley & Sons, Inc., New York.

Butler, G. M., *Handbook of Minerals*, John Wiley & Sons, Inc., New York.

Butler, G. M., and Tenney, J. B., *Petroleum*, University of Arizona, Bureau of Mines, Bulletin No. 130, 1931.

Cahen, E., and Wooten, W. O., *The Mineralogy of the Rarer Metals*, Charles Griffen and Company, Ltd. London, England, 1920.

Dake, H. C., and DeMent, Jack, *Fluorescent Light and Its Applications*, Chemical Publishing Company, Inc., Brooklyn, N.Y., 1941.

DeMent, Jack, *Fluorescent Chemicals and Their Applications*, Chemical Publishing Company, Inc.. Brooklyn, N.Y., 1942.

Ford, W. E., *Dana's Manual of Mineralogy*, John Wiley & Sons, Inc., New York, 1912.

Ladoo, R. B., *Non-Metallic Minerals*, McGraw-Hill Book Company, Inc., New York.

Marks, L. S., *Mechanical Engineers' Handbook*, McGraw-Hill Book Co., Inc., New York.

Morrison, R. S., and DeSoto, E. D., *Mining Rights on the Public Domain*, Smith-Brooks Printing Co., Denver, Colo.

Peele, R., *Mining Engineers' Handbook*, John Wiley & Sons, Inc., New York.

Prescott, A. B., and Johnson, O. C., *Qualitative Chemical Analysis*, D. Van Nostrand Company, New York.

Rastall, R. H., *Molybdenum Ores*, John Murray, London, England, 1922.

Ricketts, A. H., *American Mining Law, Division of Mines, Department of Natural Resources*, State of California. Bull. No. 98, 1931.

Rogers, A. F., *Introduction to the Study of Minerals*, McGraw-Hill Book Co., Inc., New York.

Scott, W. W., *Standard Methods of Chemical Analysis*, D. Van Nostrand Co., New York.

Smith and Miller, *An Introduction to Qualitative Analysis*, McGraw-Hill Book Co., Inc., New York, pp. 44, 259.

Struckmeyer, F. C., *The Revised Code of Arizona*, Bancroft-Whitney Co., Phoenix, Ariz., 1928.

Treadwell, F. P., and Hall, W. T., *Analytical Chemistry*, John Wiley & Sons, Inc., New York.

Warren, C. H., *Manual of Determinative Mineralogy*, McGraw-Hill Book Co., New York, 1921.

Warren, Thomas S., "White Magic with Black Light," *The Mining Journal*, Phoenix, Ariz., October 30, 1940.

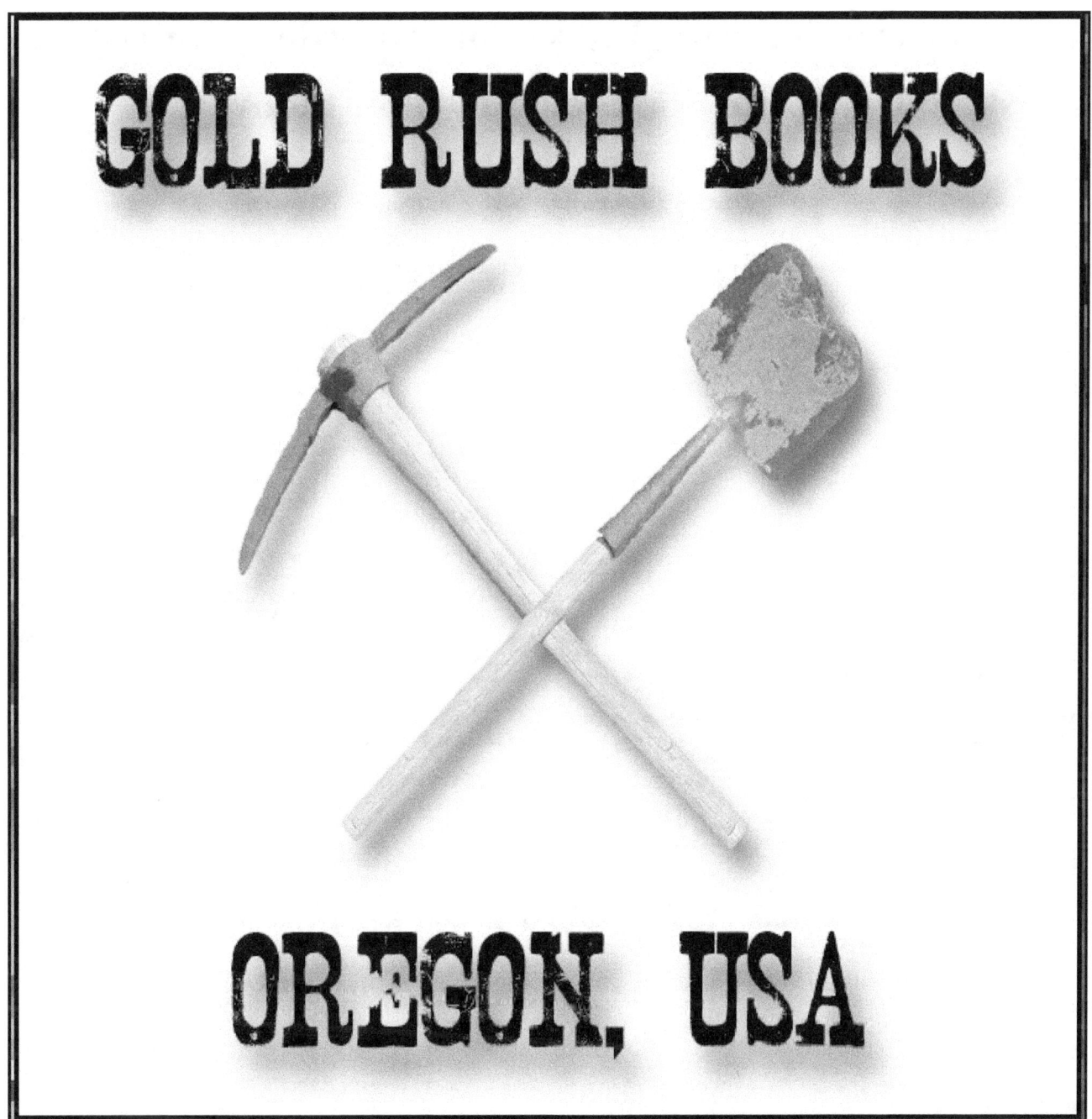

www.GoldMiningBooks.com

Books On Mining

Visit: www.goldminingbooks.com to order your copies or ask your favorite book seller to offer them.

Mining Books by Kerby Jackson

Gold Dust: Stories From Oregon's Mining Years - Oregon mining historian and prospector, Kerby Jackson, brings you a treasure trove of seventeen stories on Southern Oregon's rich history of gold prospecting, the prospectors and their discoveries, and the breathtaking areas they settled in and made homes. 5" X 8", 98 ppgs. Retail Price: $11.99

The Golden Trail: More Stories From Oregon's Mining Years - In his follow-up to "Gold Dust: Stories of Oregon's Mining Years", this time around, Jackson brings us twelve tales from Oregon's Gold Rush, including the story about the first gold strike on Canyon Creek in Grant County, about the old timers who found gold by the pail full at the Victor Mine near Galice, how Iradel Bray discovered a rich ledge of gold on the Coquille River during the height of the Rogue River War, a tale of two elderly miners on the hunt for a lost mine in the Cascade Mountains, details about the discovery of the famous Armstrong Nugget and others. 5" X 8", 70 ppgs. Retail Price: $10.99

Oregon Mining Books

Geology and Mineral Resources of Josephine County, Oregon - Unavailable since the 1970's, this important publication was originally compiled by the Oregon Department of Geology and Mineral Industries and includes important details on the economic geology and mineral resources of this important mining area in South Western Oregon. Included are notes on the history, geology and development of important mines, as well as insights into the mining of gold, copper, nickel, limestone, chromium and other minerals found in large quantities in Josephine County, Oregon. 8.5" X 11", 54 ppgs. Retail Price: $9.99

Mines and Prospects of the Mount Reuben Mining District - Unavailable since 1947, this important publication was originally compiled by geologist Elton Youngberg of the Oregon Department of Geology and Mineral Industries and includes detailed descriptions, histories and the geology of the Mount Reuben Mining District in Josephine County, Oregon. Included are notes on the history, geology, development and assay statistics, as well as underground maps of all the major mines and prospects in the vicinity of this much neglected mining district. 8.5" X 11", 48 ppgs. Retail Price: $9.99

The Granite Mining District - Notes on the history, geology and development of important mines in the well known Granite Mining District which is located in Grant County, Oregon. Some of the mines discussed include the Ajax, Blue Ribbon, Buffalo, Continental, Cougar-Independence, Magnolia, New York, Standard and the Tillicum. Also included are many rare maps pertaining to the mines in the area. 8.5" X 11", 48 ppgs. Retail Price: $9.99

Ore Deposits of the Takilma and Waldo Mining Districts of Josephine County, Oregon - The Waldo and Takilma mining districts are most notable for the fact that the earliest large scale mining of placer gold and copper in Oregon took place in these two areas. Included are details about some of the earliest large gold mines in the state such as the Llano de Oro, High Gravel, Cameron, Platerica, Deep Gravel and others, as well as copper mines such as the famous Queen of Bronze mine, the Waldo, Lily and Cowboy mines. This volume also includes six maps and 20 original illustrations. 8.5" X 11", 74 ppgs. Retail Price: $9.99

Metal Mines of Douglas, Coos and Curry Counties, Oregon - Oregon mining historian Kerby Jackson introduces us to a classic work on Oregon's mining history in this important re-issue of Bulletin 14C Volume 1, otherwise known as the Douglas, Coos & Curry Counties, Oregon Metal Mines Handbook. Unavailable since 1940, this important publication was originally compiled by the Oregon Department of Geology and Mineral Industries includes detailed descriptions, histories and the geology of over 250 metallic mineral mines and prospects in this rugged area of South West Oregon. 8.5" X 11", 158 ppgs. Retail Price: $19.99

Metal Mines of Jackson County, Oregon - Unavailable since 1943, this important publication was originally compiled by the Oregon Department of Geology and Mineral Industries includes detailed descriptions, histories and the geology of over 450 metallic mineral mines and prospects in Jackson County, Oregon. Included are such famous gold mining areas as Gold Hill, Jacksonville, Sterling and the Upper Applegate. **8.5" X 11", 220 ppgs. Retail Price: $24.99**

Metal Mines of Josephine County, Oregon - Oregon mining historian Kerby Jackson introduces us to a classic work on Oregon's mining history in this important re-issue of Bulletin 14C, otherwise known as the Josephine County, Oregon Metal Mines Handbook. Unavailable since 1952, this important publication was originally compiled by the Oregon Department of Geology and Mineral Industries includes detailed descriptions, histories and the geology of over 500 metallic mineral mines and prospects in Josephine County, Oregon. **8.5" X 11", 250 ppgs. Retail Price: $24.99**

Metal Mines of North East Oregon - Oregon mining historian Kerby Jackson introduces us to a classic work on Oregon's mining history in this important re-issue of Bulletin 14A and 14B, otherwise known as the North East Oregon Metal Mines Handbook. Unavailable since 1941, this important publication was originally compiled by the Oregon Department of Geology and Mineral Industries and includes detailed descriptions, histories and the geology of over 750 metallic mineral mines and prospects in North Eastern Oregon. **8.5" X 11", 310 ppgs. Retail Price: $29.99**

Metal Mines of North West Oregon - Oregon mining historian Kerby Jackson introduces us to a classic work on Oregon's mining history in this important re-issue of Bulletin 14D, otherwise known as the North West Oregon Metal Mines Handbook. Unavailable since 1951, this important publication was originally compiled by the Oregon Department of Geology and Mineral Industries and includes detailed descriptions, histories and the geology of over 250 metallic mineral mines and prospects in North Western Oregon. **8.5" X 11", 182 ppgs. Retail Price: $19.99**

Mines and Prospects of Oregon - Mining historian Kerby Jackson introduces us to a classic mining work by the Oregon Bureau of Mines in this important re-issue of The Handbook of Mines and Prospects of Oregon. Unavailable since 1916, this publication includes important insights into hundreds of gold, silver, copper, coal, limestone and other mines that operated in the State of Oregon around the turn of the 19th Century. Included are not only geological details on early mines throughout Oregon, but also insights into their history, production, locations and in some cases, also included are rare maps of their underground workings. **8.5" X 11", 314 ppgs. Retail Price: $24.99**

Lode Gold of the Klamath Mountains of Northern California and South West Oregon
(See California Mining Books)

Mineral Resources of South West Oregon - Unavailable since 1914, this publication includes important insights into dozens of mines that once operated in South West Oregon, including the famous gold fields of Josephine and Jackson Counties, as well as the Coal Mines of Coos County. Included are not only geological details on early mines throughout South West Oregon, but also insights into their history, production and locations. **8.5" X 11", 154 ppgs. Retail Price: $11.99**

Chromite Mining in The Klamath Mountains of California and Oregon
(See California Mining Books)

Southern Oregon Mineral Wealth - Unavailable since 1904, this rare publication provides a unique snapshot into the mines that were operating in the area at the time. Included are not only geological details on early mines throughout South West Oregon, but also insights into their history, production and locations. Some of the mining areas include Grave Creek, Greenback, Wolf Creek, Jump Off Joe Creek, Granite Hill, Galice, Mount Reuben, Gold Hill, Galls Creek, Kane Creek, Sardine Creek, Birdseye Creek, Evans Creek, Foots Creek, Jacksonville, Ashland, the Applegate River, Waldo, Kerby and the Illinois River, Althouse and Sucker Creek, as well as insights into local copper mining and other topics. **8.5" X 11", 64 ppgs. Retail Price: $8.99**

Geology and Ore Deposits of the Takilma and Waldo Mining Districts - Unavailable since the 1933, this publication was originally compiled by the United States Geological Survey and includes details on gold and copper mining in the Takilma and Waldo Districts of Josephine County, Oregon. The Waldo and Takilma mining districts are most notable for the fact that the earliest large scale mining of placer gold and copper in Oregon took place in these two areas. Included in this report are details about some of the earliest large gold mines in the state such as the Llano de Oro, High Gravel, Cameron, Platerica, Deep Gravel and others, as well as copper mines such as the famous Queen of Bronze mine, the Waldo, Lily and Cowboy mines. In addition to geological examinations, insights are also provided into the production, day to day operations and early histories of these mines, as well as calculations of known mineral reserves in the area. This volume also includes six maps and 20 original illustrations. **8.5" X 11", 74 ppgs. Retail Price: $9.99**

Gold Mines of Oregon - Oregon mining historian Kerby Jackson introduces us to a classic work on Oregon's mining history in this important re-issue of Bulletin 61, otherwise known as "Gold and Silver In Oregon". Unavailable since 1968, this important publication was originally compiled by geologists Howard C. Brooks and Len Ramp of the Oregon Department of Geology and Mineral Industries and includes detailed descriptions, histories and the geology of over 450 gold mines Oregon. Included are notes on the history, geology and gold production statistics of all the major mining areas in Oregon including the Klamath Mountains, the Blue Mountains and the North Cascades. While gold is where you find it, as every miner knows, the path to success is to prospect for gold where it was previously found. **8.5" X 11", 344 ppgs. Retail Price: $24.99**

Mines and Mineral Resources of Curry County Oregon - Originally published in 1916, this important publication on Oregon Mining has not been available for nearly a century. Included are rare insights into the history, production and locations of dozens of gold mines in Curry County, Oregon, as well as detailed information on important Oregon mining districts in that area such as those at Agness, Bald Face Creek, Mule Creek, Boulder Creek, China Diggings, Collier Creek, Elk River, Gold Beach, Rock Creek, Sixes River and elsewhere. Particular attention is especially paid to the famous beach gold deposits of this portion of the Oregon Coast. **8.5" X 11", 140 ppgs. Retail Price: $11.99**

Chromite Mining in South West Oregon - Originally published in 1961, this important publication on Oregon Mining has not been available for nearly a century. Included are rare insights into the history, production and locations of nearly 300 chromite mines in South Western Oregon. **8.5" X 11", 184 ppgs. Retail Price: $14.99**

Mineral Resources of Douglas County Oregon - Originally published in 1972, this important publication on Oregon Mining has not been available for nearly forty years. Included are rare insights into the geology, history, production and locations of numerous gold mines and other mining properties in Douglas County, Oregon. **8.5" X 11", 124 ppgs. Retail Price: $11.99**

Mineral Resources of Coos County Oregon - Originally published in 1972, this important publication on Oregon Mining has not been available for nearly forty years. Included are rare insights into the geology, history, production and locations of numerous gold mines and other mining properties in Coos County, Oregon. **8.5" X 11", 100 ppgs. Retail Price: $11.99**

Mineral Resources of Lane County Oregon - Originally published in 1938, this important publication on Oregon Mining has not been available for nearly seventy five years. Included are extremely rare insights into the geology and mines of Lane County, Oregon, in particular in the Bohemia, Blue River, Oakridge, Black Butte and Winberry Mining Districts. **8.5" X 11", 82 ppgs. Retail Price: $9.99**

Mineral Resources of the Upper Chetco River of Oregon: Including the Kalmiopsis Wilderness - Originally published in 1975, this important publication on Oregon Mining has not been available for nearly forty years. Withdrawn under the 1872 Mining Act since 1984, real insight into the minerals resources and mines of the Upper Chetco River has long been unavailable due to the remoteness of the area. Despite this, the decades of battle between property owners and environmental extremists over the last private mining inholding in the area has continued to pique the interest of those interested in mining and other forms of natural resource use. Gold mining began in the area in the 1850's and has a rich history in this geographic area, even if the facts surrounding it are little known. Included are twenty two rare photographs, as well as insights into the Becca and Morning Mine, the Emmly Mine (also known as Emily Camp), the Frazier Mine, the Golden Dream or Higgins Mine, Hustis Mine, Peck Mine and others. **8.5" X 11", 64 ppgs. Retail Price: $8.99**

Gold Dredging in Oregon - Originally published in 1939, this important publication on Oregon Mining has not been available for nearly seventy five years. Included are extremely rare insights into the history and day to day operations of the dragline and bucketline gold dredges that once worked the placer gold fields of South West and North East Oregon in decades gone by. Also included are details into the areas that were worked by gold dredges in Josephine, Jackson, Baker and Grant counties, as well as the economic factors that impacted this mining method. This volume also offers a unique look into the values of river bottom land in relation to both farming and mining, in how farm lands were mined, re-soiled and reclamated after the dredges worked them. Featured are hard to find maps of the gold dredge fields, as well as rare photographs from a bygone era. **8.5" X 11", 86 ppgs. Retail Price: $8.99**

Quick Silver Mining in Oregon - Originally published in 1963, this important publication on Oregon Mining has not been available for over fifty years. This publication includes details into the history and production of Elemental Mercury or Quicksilver in the State of Oregon. **8.5" X 11", 238 ppgs. Retail Price: $15.99**

Mines of the Greenhorn Mining District of Grant County Oregon - Originally published in 1948, this important publication on Oregon Mining has not been available for over sixty five years. In this publication are rare insights into the mines of the famous Greenhorn Mining District of Grant County, Oregon, especially the famous Morning Mine. Also included are details on the Tempest, Tiger, Bi-Metallic, Windsor, Psyche, Big Johnny, Snow Creek, Banzette and Paramount Mines, as well as prospects in the vicinities in the famous mining areas of Mormon Basin, Vinegar Basin and Desolation Creek. Included are hard to find mine maps and dozens of rare photographs from the bygone era of Grant County's rich mining history. **8.5" X 11", 72 ppgs. Retail Price: $9.99**

Geology of the Wallowa Mountains of Oregon: Part I (Volume 1) - Originally published in 1938, this important publication on Oregon Mining has not been available for nearly seventy five years. Included are details on the geology of this unique portion of North Eastern Oregon. This is the first part of a two book series on the area. Accompanying the text are rare photographs and historic maps.**8.5" X 11", 92 ppgs. Retail Price: $9.99**

Geology of the Wallowa Mountains of Oregon: Part II (Volume 2) - Originally published in 1938, this important publication on Oregon Mining has not been available for nearly seventy five years. Included are details on the geology of this unique portion of North Eastern Oregon. This is the first part of a two book series on the area. Accompanying the text are rare photographs and historic maps.**8.5" X 11", 94 ppgs. Retail Price: $9.99**

Field Identification of Minerals For Oregon Prospectors - Originally published in 1940, this important publication on Oregon Mining has not been available for nearly seventy five years. Included in this volume is an easy system for testing and identifying a wide range of minerals that might be found by prospectors, geologists and rockhounds in the State of Oregon, as well as in other locales. Topics include how to put together your own field testing kit and how to conduct rudimentary tests in the field. This volume is written in a clear and concise way to make it useful even for beginners. **8.5" X 11", 158 ppgs. Retail Price: $14.99**

The Bohemia Mining District of Oregon - Originally published in 1900, this important publication on Oregon Mining has not been available for over a century. Included in this volume are important insights into the famous Bohemia Mining District of Oregon, including the histories and locations of important gold mines in the area such as the Ophir Mine, Clarence, Acturas, Peek-a-boo, White Swan, Combination Mine, the Musick Mine, The California, White Ghost, The Mystery, Wall Street, Vesuvius, Story, Lizzie Bullock, Delta, Elsie Dora, Golden Slipper, Broadway, Champion Mine, Knott, Noonday, Helena, White Wings, Riverside and others. Also included are notes on the nearby Blue River Mining District. **8.5" X 11", 58 ppgs. Retail Price: $9.99**

The Gold Fields of Eastern Oregon - Unavailable since 1900, this publication was originally compiled by the Baker City Chamber of Commerce Offering important insights into the gold mining history of Eastern Oregon, "The Gold Fields of Eastern Oregon" sheds a rare light on many of the gold mines that were operating at the turn of the 19th Century in Baker County and Grant County in North Eastern Oregon. Some of the areas featured include the Cable Cove District, Baisely-Elhorn, Granite, Red Boy, Bonanza, Susanville, Sparta, Virtue, Vaughn, Sumpter, Burnt River, Rye Valley and other mining districts. Included is basic information on not only many gold mines that are well known to those interested in Eastern Oregon mining history, but also many mines and prospects which have been mostly lost to the passage of time. Accompanying are numerous rare photos **8.5" X 11", 78 ppgs. Retail Price: $10.99**

Gold Mining in Eastern Oregon - Originally published in 1938, this important publication on Oregon Mining has not been available for over a century. Included in this volume are important insights into the famous mining districts of Eastern Oregon during the late 1930's. Particular attention is given to those gold mines with milling and concentrating facilities in the Greenhorn, Red Boy, Alamo, Bonanza, Granite, Cable Cove, Cracker Creek, Virtue, Keating, Medical Springs, Sanger, Sparta, Chicken Creek, Mormon Basin, Connor Creek, Cornucopia and the Bull Run Mining Districts. Some of the mines featured include the Ben Harrison, North Pole-Columbia, Highland Maxwell, Baisley-Elkhorn, White Swan, Balm Creek, Twin Baby, Gem of Sparta, New Deal, Gleason, Gifford-Johnson, Cornucopia, Record, Bull Run, Orion and others. Of particular interest are the mill flow sheets and descriptions of milling operations of these mines. **8.5" X 11", 68 ppgs. Retail Price: $8.99**

The Gold Belt of the Blue Mountains of Oregon - Originally published in 1901, this important publication on Oregon Mining has not been available for over a century. Included in this volume are rare insights into the gold deposits of the Blue Mountains of North East Oregon, including the history of their early discovery and early production. Extensive details are offered on this important mining area's mineralogy and economic geology, as well as insights into nearby gold placers, silver deposits and copper deposits. Featured are the Elkhorn and Rock Creek mining districts, the Pocahontas district, Auburn and Minersville districts, Sumpter and Cracker Creek, Cable Cove, the Camp Carson district, Granite, Alamo, Greenhorn, Robinsonville, the Upper Burnt River Valley and Bonanza districts, Susanville, Quartzburg, Canyon Creek, Virtue, the Copper Butte district, the North Powder River, Sparta, Eagle Creek, Cornucopia, Pine Creek, Lower Powder River, the Upper Snake River Canyon, Rye Valley, Lower Burnt River Valley, Mormon Basin, the Malheur and Clarks Creek districts, Sutton Creek and others. Of particular interest are important details on numerous gold mines and prospects in these mining districts, including their locations, histories, geology and other important information, as well as information on silver, copper and fire opal deposits. **8.5" X 11", 250 ppgs. Retail Price: $24.99**

<u>Mining in the Cascades Range of Oregon</u> - Originally published in 1938, this important publication on Oregon Mining has not been available for over seventy five years. Included in this volume are rare insights into the gold mines and other types of metal mines in the Cascades Mountain Range of Oregon. Some of the important mining areas covered include the famous Bohemia Mining District, the North Santiam Mining District, Quartzville Mining District, Blue River Mining District, Fall Creek Mining District, Oakridge District, Zinc District, Buzzard-Al Sarena District, Grand Cove, Climax District and Barron Mining District. Of particular interest are important details on over 100 mines and prospects in these mining districts, including their locations, histories, geology and other important information. **8.5" X 11", 170 ppgs. Retail Price: $14.99**

<u>Beach Gold Placers of the Oregon Coast</u> - Originally published in 1934, this important publication on Oregon Mining has not been available for over 80 years. Included in this volume are rare insights into the beach gold deposits of the State of Oregon, including their locations, occurance, composition and geology. Of particular interest is information on placer platinum in Oregon's rich beach deposits. Also included are the locations and other information on some famous Oregon beach mines, including the Pioneer, Eagle, Chickamin, Iowa and beach placer mines north of the mouth of the Rogue River. **8.5" X 11", 60 ppgs. Retail Price: $8.99**

<u>Mineralogical Composition of the Sands of the Oregon Coast: From Coos Bay to the Columbia</u> - Published in 1945, he text features hard to find information on the composition of the gold bearing black sands of the South West Oregon Coast, offering a unique insight to prospectors in search of Oregon's legendary beach gold. 104 ppgs, $9.99

<u>Manganese Mining in Oregon</u> - First released in 1942 and now out of print, this special reprint edition of "Manganese in Oregon" was originally published by the Oregon Department of Geology and Mineral Industries. The text features hard to find information on the mining of Manganese in Oregon, including details and maps of Oregon manganese mines and prospects. 108 ppgs, 9.99

<u>Medford Oregon As A Mining Center</u> - Written in 1912, this hard to find publication includes valuable insights into the mining history of South West Oregon. This small book contains interesting information on the gold, copper and mining industry in Southern Oregon as it existed just prior to World War One, shedding light on some of the important mines in the area. Included are rare photographs and vintage advertising of the day. 80 ppgs, 9.99

<u>Mineral Resources of Curry County Oregon</u> - First released in 1977 and now out of print, this special reprint edition of "Geology, Mineral Resources and Rock Materials of Curry County, Oregon" was originally published in cooperation of Curry County, Oregon and the Oregon Department of Geology and Mineral Industries. The text features hard to find information on not only the mining of gold and other metals in Curry County, but also aggregate mining in the area. 102 ppgs, 11.99

<u>Origin of the Gold Bearing Black Sands of the Coast of South West Oregon</u> - First released in 1943 and now out of print, this special reprint edition of "The Origin of the Black Sands of the South West Oregon Coast" was originally published by the Oregon Department of Geology and Mineral Industries. The text features hard to find information on the origin of the gold bearing black sands of the South West Oregon Coast, offering a unique insight to prospectors in search of Oregon's legendary beach gold. 52 ppgs, 8.99

<u>South West Oregon Mining</u> - Leading mining historian Kerby Jackson introduces us to six classic small mining publications on the Gold Mining Industry in Southern Oregon. This small book consists of a compilation of USGS J.S. Diller's "Mines of the Riddles Quadrangle", "The Rogue River Valley Coal Fields" and "Mineral Resources of the Grants Pass Quadrangle", the Grants Pass Commercial Club's rare publication "Mining in Josephine County, Oregon" and the USGS publication "The Distribution of Placer Gold in the Sixes River, South West Oregon". Also included is F.W. Libbey's legendary article on the Southern Oregon Mining Industry, "Lest We Forget", which appeared in the publication of the Oregon State Department of Geology and Mineral Industries in the early 1960's. This compilation offers a unique perspective on mining in South West Oregon and includes considerable information on mines in Josephine, Jackson and Coos Counties. 142 ppgs, 14.99

<u>Geology and Mineral Resources of the Gasquet Quadrangle of California-Oregon</u> - First published in 1953, it has been unavailable for over a century and sheds important light on the geological features and mineral resources of this portion of Northern California and Southern Oregon. 80 ppgs, 9.99

Idaho Mining Books

Gold in Idaho - Unavailable since the 1940's, this publication was originally compiled by the Idaho Bureau of Mines and includes details on gold mining in Idaho. Included is not only raw data on gold production in Idaho, but also valuable insight into where gold may be found in Idaho, as well as practical information on the gold bearing rocks and other geological features that will assist those looking for placer and lode gold in the State of Idaho. This volume also includes thirteen gold maps that greatly enhance the practical usability of the information contained in this small book detailing where to find gold in Idaho. **8.5" X 11", 72 ppgs. Retail Price: $9.99**

Geology of the Couer D'Alene Mining District of Idaho - Unavailable since 1961, this publication was originally compiled by the Idaho Bureau of Mines and Geology and includes details on the mining of gold, silver and other minerals in the famous Coeur D'Alene Mining District in Northern Idaho. Included are details on the early history of the Coeur D'Alene Mining District, local tectonic settings, ore deposit features, information on the mineral belts of the Osburn Fault, as well as detailed information on the famous Bunker Hill Mine, the Dayrock Mine, Galena Mine, Lucky Friday Mine and the infamous Sunshine Mine. This volume also includes sixteen hard to find maps. **8.5" X 11", 70 ppgs. Retail Price: $9.99**

The Gold Camps and Silver Cities of Idaho - Originally published in 1963, this important publication on Idaho Mining has not been available for nearly fifty years. Included are rare insights into the history of Idaho's Gold Rush, as well as the mad craze for silver in the Idaho Panhandle. Documented in fine detail are the early mining excitements at Boise Basin, at South Boise, in the Owyhees, at Deadwood, Long Valley, Stanley Basin and Robinson Bar, at Atlanta, on the famous Boise River, Volcano, Little Smokey, Banner, Boise Ridge, Hailey, Leesburg, Lemhi, Pearl, at South Mountain, Shoup and Ulysses, Yellow Jacket and Loon Creek. The story follows with the appearance of Chinese miners at the new mining camps on the Snake River, Black Pine, Yankee Fork, Bay Horse, Clayton, Heath, Seven Devils, Gibbonsville, Vienna and Sawtooth City. Also included are special sections on the Idaho Lead and Silver mines of the late 1800's, as well as the mining discoveries of the early 1900's that paved the way for Idaho's modern mining and mineral industry. Lavishly illustrated with rare historic photos, this volume provides a one of a kind documentary into Idaho's mining history that is sure to be enjoyed by not only modern miners and prospectors who still scour the hills in search of nature's treasures, but also those enjoy history and tromping through overgrown ghost towns and long abandoned mining camps. **8.5" X 11", 186 ppgs. Retail Price: $14.99**

Ore Deposits and Mining in North Western Custer County Idaho - Unavailable since 1913, this important publication was originally published by the Us Department of the Interior and has been unavailable for a century. Included are fine details on the geology, geography, gold placers and gold and silver bearing quartz veins of the mining region of North West Custer County, Idaho. Of particular interest is a rare look at the mines and prospects of the region, including those such as the Ramshorn Mine, SkyLark, Riverview, Excelsior, Beardsley, Pacific, Hoosier, Silver Brick, Forest Rose and dozens of others in the Bay Horse Mining District. Also covered are the mines of the Yankee Fork District such as the Lucky Boy, Badger, Black, Enterprise, Charles Dickens, Morrison, Golden Sunbeam, Montana, Golden Gate and others, as well as those in the Loon Mining District. **8.5" X 11", 126 ppgs. Retail Price: $12.99**

Gold Rush To Idaho - Unavailable since 1963, this important publication was originally published by the Idaho Bureau of Mines and has been unavailable for 50 years. "Gold Rush To Idaho" revisits the earliest years of the discovery of gold in Idaho Territory and introduces us to the conditions that the pioneer gold seekers met when they blazed a trail through the wilderness of Idaho's mountains and discovered the precious yellow metal at Oro Fino and Pierce. Subsequent rushes followed at places like Elk City, Newsome, Clearwater Station, Florence, Warrens and elsewhere. Of particular interest is a rare look at the hardships that the first miners in Idaho met with during their day to day existences and their attempts to bring law and order to their mining camps. **8.5" X 11", 88 ppgs. Retail Price: $9.99**

The Geology and Mines of Northern Idaho and North Western Montana - Unavailable since 1909, this important publication was originally published by the Us Department of the Interior and has been unavailable for a century. Included are fine details on the geology and geography of the mining regions of Northern Idaho and North Western Montana. Of particular interest is a rare look at the mines and prospects of the region, including those in the Pine Creek Mining District, Lake Pend Oreille district, Troy Mining District, Sylvanite District, Cabinet Mining District, Prospect Mining District and the Missoula Valley. Some of the mines featured include the Iron Mountain, Silver Butte, Snowshoe, Grouse Mountain Mine and others. **8.5" X 11", 142 ppgs. Retail Price: $12.99**

Mining in the Alturas Quadrangle of Blaine County Idaho - Unavailable since 1922, this important publication was originally published by the Idaho Bureau of Mines and has been unavailable for ninety years. Topics include the geology, rock formations and the formation of ore deposits in this important mining area of Idaho. Of particular focus is information on the local geology, quartz veins and ore deposits of this portion of Idaho. Included are hard to find details, including the descriptions and locations of numerous gold and silver mines in the area including the Silver King, Pilgrim, Columbia, Lone Jack, Sunbeam, Pride of the West, Lucky Boy, Scotia, Atlanta, Beaver-Bidwell and others mines and prospects. **8.5" X 11", 56 ppgs. Retail Price: $8.99**

Mining in Lemhi County Idaho - Originally published in 1913, this important book on Idaho Mining has not been available to miners for over a century. Included are rare insights into hundreds of gold, silver, copper and other mines in this famous Idaho mining area. Details include the locations, geology, history, production and other facts of the mines of this region, not only gold and silver hardrock mines, but also gold placer mines, lead-silver deposits, copper mines, cobalt-nickel deposits, tungsten and tin mines . It is lavishly illustrated with hard to find photos of the period and rare mining maps. Some of the vicinities featured include the Nicholia Mining District, Spring Mountain District, Texas District, Blue Wing District, Junction District, McDevitt District, Pratt Creek, Eldorado District, Kirtley Creek, Carmen Creek, Gibbonsville, Indian Creek, Mineral Hill District, Mackinaw, Eureka District, Blackbird District, YellowJacket District, Gravel Range District, Junction District, Parker Mountain and other mining districts. 8.5" X 11", 226 ppgs. Retail Price: $19.99

Mining in Shoshone County Idaho - First published in 1923, it has been unavailable for over a century and sheds important light on the mining history of Shoshone County, Idaho. Some of the topics include the history of mining in Shoshone County, a look at the local geology and ore characteristics of lead-silver deposits, zinc deposits, copper, antimony, gold and other minerals. Also included are insights into the history, production, characteristics and locations of numerous mines in the area. 198 ppgs, 15.99

Utah Mining Books

Fluorite in Utah - Unavailable since 1954, this publication was originally compiled by the USGS, State of Utah and U.S. Atomic Energy Commission and details the mining of fluorspar, also known as fluorite in the State of Utah. Included are details on the geology and history of fluorspar (fluorite) mining in Utah, including details on where this unique gem mineral may be found in the State of Utah. 8.5" X 11", 60 ppgs. Retail Price: $8.99

The Gold Hill Mining District of Utah - First published in 1935, it has been unavailable since those days and sheds important light on the mines, history and geology of Utah's Gold Hill Mining District. Included are rare insights into this important mining area, including the locations, histories and details of numerous mines. This volume is well illustrated with geological diagrams, as well as hard to find maps of some of the most important mines in this district. 202 ppgs., 19.99

The Mines, Miners and Minerals of Utah - First published in 1896, it has been unavailable since those days and sheds important light on the early mines and miners of Pioneer Utah, as well as the minerals which they won from the earth by laborious hard physical labor and sheer determination. Included are rare insights into the early mining history of Utah, as well details on hundreds of gold, silver and copper mines. 376 ppgs., 24.99

California Mining Books

The Tertiary Gravels of the Sierra Nevada of California - Mining historian Kerby Jackson introduces us to a classic mining work by Waldemar Lindgren in this important re-issue of The Tertiary Gravels of the Sierra Nevada of California. Unavailable since 1911, this publication includes details on the gold bearing ancient river channels of the famous Sierra Nevada region of California. 8.5" X 11", 282 ppgs. Retail Price: $19.99

The Mother Lode Mining Region of California - Unavailable since 1900, this publication includes details on the gold mines of California's famous Mother Lode gold mining area. Included are details on the geology, history and important gold mines of the region, as well as insights into historic mining methods, mine timbering, mining machinery, mining bell signals and other details on how these mines operated. Also included are insights into the gold mines of the California Mother Lode that were in operation during the first sixty years of California's mining history. 8.5" X 11", 176 ppgs. Retail Price: $14.99

Lode Gold of the Klamath Mountains of Northern California and South West Oregon - Unavailable since 1971, this publication was originally compiled by Preston E. Hotz and includes details on the lode mining districts of Oregon and California's Klamath Mountains. Included are details on the geology, history and important lode mines of the French Gulch, Deadwood, Whiskeytown, Shasta, Redding, Muletown, South Fork, Old Diggings, Dog Creek (Delta), Bully Choop (Indian Creek), Harrison Gulch, Hayfork, Minersville, Trinity Center, Canyon Creek, East Fork, New River, Denny, Liberty (Black Bear), Cecilville, Callahan, Yreka, Fort Jones and Happy Camp mining districts in California, as well as the Ashland, Rogue River, Applegate, Illinois River, Takilma, Greenback, Galice, Silver Peak, Myrtle Creek and Mule Creek districts of South Western Oregon. Also included are insights into the mineralization and other characteristics of this important mining region. 8.5" X 11", 100 ppgs. Retail Price: $10.99

Mines and Mineral Resources of Shasta County, Siskiyou County, Trinity County: California - Unavailable since 1915, this publication was originally compiled by the California State Mining Bureau and includes details on the gold mines of this area of Northern California. Also included are insights into the mineralization and other characteristics of this important mining region, as well as the location of historic gold mines. 8.5" X 11", 204 ppgs. Retail Price: $19.99

Geology of the Yreka Quadrangle, Siskiyou County, California - Unavailable since 1977, this publication was originally compiled by Preston E. Hotz and includes details on the geology of the Yreka Quadrangle of Siskiyou County, California. Also included are insights into the mineralization and other characteristics of this important mining region. **8.5" X 11", 78 ppgs. Retail Price: $7.99**

Mines of San Diego and Imperial Counties, California - Originally published in 1914, this important publication on California Mining has not been available for a century. This publication includes important information on the early gold mines of San Diego and Imperial County, which were some of the first gold fields mined in California by early Spanish and Mexican miners before the 49ers came on the scene. Included are not only details on early mining methods in the area, production statistics and geological information, but also the location of the early gold mines that helped make California "The Golden State". Also included are details on the mining of other minerals such as silver, lead, zinc, manganese, tungsten, vanadium, asbestos, barite, borax, cement, clay, dolomite, fluospar, gem stones, graphite, marble, salines, petroleum, stronium, talc and others. **8.5" X 11", 116 ppgs. Retail Price: $12.99**

Mines of Sierra County, California - Unavailable since 1920, this publication was originally compiled by the California State Mining Bureau and includes details on the gold mines of Sierra County, California. Also included are insights into the mineralization and other characteristics of this important mining region, as well as the location of historic gold mines. **8.5" X 11", 156 ppgs. Retail Price: $19.99**

Mines of Plumas County, California - Unavailable since 1918, this publication was originally compiled by the California State Mining Bureau and includes details on the gold mines of Plumas County, California. Also included are insights into the mineralization and other characteristics of this important mining region, as well as the location of historic gold mines. **8.5" X 11", 200 ppgs. Retail Price: $19.99**

Mines of El Dorado, Placer, Sacramento and Yuba Counties, California - Originally published in 1917, this important publication on California Mining has not been available for nearly a century. This publication includes important information on the early gold mines of El Dorado County, Placer County, Sacramento County and Yuba County, which were some of the first gold fields mined by the Forty-Niners during the California Gold Rush. Included are not only details on early mining methods in the area, production statistics and geological information, but also the location of the early gold mines that helped make California "The Golden State". Also included are insights into the early mining of chrome, copper and other minerals in this important mining area. **8.5" X 11", 204 ppgs. Retail Price: $19.99**

Mines of Los Angeles, Orange and Riverside Counties, California - Originally published in 1917, this important publication on California Mining has not been available for nearly a century. This publication includes important information on the early gold mines of Los Angeles County, Orange County and Riverside County, which were some of the first gold fields mined in California by early Spanish and Mexican miners before the 49ers came on the scene. Included are not only details on early mining methods in the area, production statistics and geological information, but also the location of the early gold mines that helped make California "The Golden State". **8.5" X 11", 146 ppgs. Retail Price: $12.99**

Mines of San Bernadino and Tulare Counties, California - Originally published in 1917, this important publication on California Mining has not been available for nearly a century. This publication includes important information on the early gold mines of San Bernadino and Tulare County, which were some of the first gold fields mined in California by early Spanish and Mexican miners before the 49ers came on the scene. Included are not only details on early mining methods in the area, production statistics and geological information, but also the location of the early gold mines that helped make California "The Golden State". Also included are details on the mining of other minerals such as copper, iron, lead, zinc, manganese, tungsten, vanadium, asbestos, barite, borax, cement, clay, dolomite, fluospar, gem stones, graphite, marble, salines, petroleum, stronium, talc and others. **8.5" X 11", 200 ppgs. Retail Price: $19.99**

Chromite Mining in The Klamath Mountains of California and Oregon - Unavailable since 1919, this publication was originally compiled by J.S. Diller of the United States Department of Geological Survey and includes details on the chromite mines of this area of Northern California and Southern Oregon. Also included are insights into the mineralization and other characteristics of this important mining region, as well as the location of historic mines. Also included are insights into chromite mining in Eastern Oregon and Montana. **8.5" X 11", 98 ppgs. Retail Price: $9.99**

Mines and Mining in Amador, Calaveras and Tuolumne Counties, California - Unavailable since 1915, this publication was originally compiled by William Tucker and includes details on the mines and mineral resources of this important California mining area. Included are details on the geology, history and important gold mines of the region, as well as insights into other local mineral resources such as asbestos, clay, copper, talc, limestone and others. Also included are insights into the mineralization and other characteristics of this important portion of California's Mother Lode mining region. **8.5" X 11", 198 ppgs. Retail Price: $14.99**

The Cerro Gordo Mining District of Inyo County California - Unavailable since 1963, this publication was originally compiled by the United States Department of Interior. Included are insights into the mineralization and other characteristics of this important mining region of Southern California. Topics include the mining of gold and silver in this important mining district in Inyo County, California, including details on the history, production and locations of the Cerro Gordo Mine, the Morning Star Mine, Estelle Tunnel, Charles Lease Tunnel, Ignacio, Hart, Crosscut Tunnel, Sunset, Upper Newtown, Newtown, Ella, Perseverance, Newsboy, Belmont and other silver and gold mines in the Cerro Gordo Mining District. This volume also includes important insights into the fossil record, geologic formations, faults and other aspects of economic geology in this California mining district. 8.5" X 11", 104 ppgs. **Retail Price: $10.99**

Mining in Butte, Lassen, Modoc, Sutter and Tehama Counties of California - Unavailable since 1917, this publication was originally compiled by the United States Department of Interior. Included are insights into the mineralization and other characteristics of this important mining region of California. Topics include the mining of asbestos, chromite, gold, diamonds and manganese in Butte County, the mining of gold and copper in the Hayden Hill and Diamond Mountain mining districts of Lassen County, the mining of coal, salt, copper and gold in the High Grade and Winters mining districts of Modoc County, gold mining in Sutter County and the mining of gold, chromite, manganese and copper in Tehama County. This volume also includes the production records and locations of numerous mines in this important mining region. 8.5" X 11", 114 ppgs. **Retail Price: $11.99**

Mines of Trinity County California - Originally published in 1965, this important publication on California Mining has not been available for nearly fifty years. This publication includes important information on mines and mining in Trinity County, California, as well insights into the mineralization and geology of this important mining area in Northern California. Included are extensive details on hardrock and placer gold mines and prospects, including charts showing the locations of these historic mines.. 8.5" X 11", 144 ppgs. **Retail Price: $12.99**

Mines of Kern County California - Originally published in 1962, this important publication on California Mining has not been available for nearly fifty years. This publication includes important information on mines and mining in Kern County, California, as well insights into the mineralization and geology of this important mining area in California. Included are extensive details on hardrock and placer gold mines and prospects, including charts showing the locations of these historic mines. 8.5" X 11", 398 ppgs. **Retail Price: $24.99**

Mines of Calaveras County California - Originally published in 1962, this important publication on California Mining has not been available for nearly fifty years. This publication includes important information on mines and mining in Calaveras County, California, as well insights into the mineralization and geology of this important mining area in Northern California. Included are extensive details on hardrock and placer gold mines and prospects, including charts showing the locations of these historic mines. 8.5" X 11", 236 ppgs. **Retail Price: $19.99**

Lode Gold Mining in Grass Valley California - Unavailable since 1940, this publication was originally compiled by the United States Department of Interior. Included are insights into the gold mineralization and other characteristics of this important mining region of Nevada County, California. This volume also includes important insights into the geologic formations, faults and other aspects of economic geology in this California mining district. Of particular interest are the fine details on many hardrock gold mines in the area, including their locations, histories, development and mineralization. Some of the mines featured include the Gold Hill Mine, Massachusetts Hill, Boundary, Peabody, Golden Center, North Star, Omaha, Lone Jack, Homeward Bound, Hartery, Wisconsin, Allison Ranch, Phoenix, Kate Hayes, W.Y.O.D., Empire, Rich Hill, Daisy Hill, Orleans, Sultana, Centennial, Conlin, Ben Franklin, Crown Point and many others. 8.5" X 11", 148 ppgs. **Retail Price: $12.99**

Lode Mining in the Alleghany District of Sierra County California - Unavailable since 1913, this publication was originally compiled by the United States Department of Interior. Included are insights into the mineralization and other characteristics of this important mining region of Sierra County. Included are details on the history, production and locations of numerous hardrock gold mines in this famous California area, including the Tightner Mine, Minnie D., Osceola, Eldorado, Twenty One, Sherman, Kenton, Oriental, Rainbow, Plumbago, Irelan, Gold Canyon, North Fork, Federal, Kate Hardy and others. This volume also includes important insights into the fossil record, geologic formations, faults and other aspects of economic geology in this California mining district. 8.5" X 11", 48 ppgs. **Retail Price: $7.99**

Six Months In The Gold Mines During The California Gold Rush - Unavailable since 1850, this important work is a first hand account of one "49'ers" personal experience during the great California Gold Rush, shedding important light on one of the most exciting periods in the history of not only California, but also the world. Compiled from journals written between 1847 and 1849 by E. Gould Buffum, a native of New York, "Six Months In The Gold Mines During The California Gold Rush" offers a rare look into the day to day lives of the people who came to California to work in her gold mines when the state was still a great frontier. 8.5" X 11", 290 ppgs. **Retail Price: $19.99**

<u>**Quartz Mines of the Grass Valley Mining District of California**</u> - Unavailable since 1867, this important publication has not been available since those days. This rare publication offers a short dissertation on the early hardrock mines in this important mining district in the California Mother Lode region between the 1850's and 1860's. Also included are hard to find details on the mineralization and locations of these mines, as well as how they were operated in those day. **8.5″ X 11″, 44 ppgs.** Retail Price: $8.99

<u>**Gold Rush on the Feather River**</u> - First published in 1924, this short publication by G.C. Mansfield sheds important light on the early history of gold mining on the Feather River. Included are rare insights into the first decade of gold mining and the early mining camps of the Feather River during the 1850's. 64 ppgs., 9.99

<u>**The Bodie Mining District of California**</u> - First published in 1986, it has been unavailable since those days and sheds important light on this famous mining area. Included are the history, characteristics and locations of numerous old mines around the ghost town of Bodie.
64 ppgs, 8.99

<u>**Geology and Mineral Resources of the Gasquet Quadrangle of California-Oregon**</u> - First published in 1953, it has been unavailable for over a century and sheds important light on the geological features and mineral resources of this portion of Northern California and Southern Oregon.
80 ppgs, 9.99

Alaska Mining Books

<u>**Ore Deposits of the Willow Creek Mining District, Alaska**</u> - Unavailable since 1954, this hard to find publication includes valuable insights into the Willow Creek Mining District near Hatcher Pass in Alaska. The publication includes insights into the history, geology and locations of the well known mines in the area, including the Gold Cord, Independence, Fern, Mabel, Lonesome, Snowbird, Schroff-O'Neil, High Grade, Marion Twin, Thorpe, Webfoot, Kelly-Willow, Lane, Holland and others. **8.5″ X 11″, 96 ppgs.** Retail Price: $9.99

<u>**The Juneau Gold Belt of Alaska**</u> - Unavailable since 1906, this hard to find publication includes valuable insights into the gold mines around Juneau, Alaska. The publication includes important details into the history, geology and locations of the well known gold mines and prospects in the area, including those around Windham Bay, Holkham Bay, Port Snettisham, on Grindstone and Rhine Creeks, Gold Creek, Douglas Island, Salmon Creek, Lemon Creek, Nugget Creek, from the Mendenhall River to Berners Bay, McGinnis Creek, Montana Creek, Peterson Creek, Windfall Creek, the Eagle River, Yankee Basin, Yankee Curve, Kowee Creek and elsewhere. Not only are gold placer mines included, but also hardrock gold mines. **8.5″ X 11″, 224 ppgs.** Retail Price: $19.99

<u>**Mining in the Jumbo Basin of Alaska**</u> - Unavailable since 1953, this hard to find publication includes valuable insights into the mines and geology of the Jumbo Basin. The publication includes important details into the history, geology and locations of the well known gold mines and prospects in the famous Jumbo Basin Mining Region of Alaska.
72 ppgs, 9.99

<u>**The Rampart Placer Gold Region of Alaska**</u> - Unavailable since 1906, this hard to find publication includes valuable insights into the placer gold mines of the Rampart Mining Region. The publication includes important details into the history, geology and locations of the well known gold mines and prospects in the famous Rampart Mining Region of Alaska.
78 ppgs, 10.99

Arizona Mining Books

<u>**Mines and Mining in Northern Yuma County Arizona**</u> - Originally published in 1911, this important publication on Arizona Mining has not been available for over a hundred years. Included are rare insights into the gold, silver, copper and quicksilver mines of Yuma County, Arizona together with hard to find maps and photographs. Some of the mines and mining districts featured include the Planet Copper Mine, Mineral Hill, the Clara Consolidated Mine, Viati Mine, Copper Basin prospect, Bowman Mine, Quartz King, Billy Mack, Carnation, the Wardwell and Osbourne, Valensuella Copper, the Mariquita, Colonial Mine, the French American, the New York-Plomosa, Guadalupe, Lead Camp, Mudersbach Copper Camp, Yellow Bird, the Arizona Northern (Salome Strike), Bonanza (Harqua Hala), Golden Eagle, Hercules, Socorro and others. **8.5″ X 11″, 144 ppgs.** Retail Price: $11.99

<u>**The Aravaipa and Stanley Mining Districts of Graham County Arizona**</u> - Originally published in 1925, this important publication on Arizona Mining has not been available for nearly ninety years. Included are rare insights into the gold and silver mines of these two important mining districts, together with hard to find maps. **8.5″ X 11″, 140 ppgs.** Retail Price: $11.99

Gold in the Gold Basin and Lost Basin Mining Districts of Mohave County, Arizona - This volume contains rare insights into the geology and gold mineralization of the Gold Basin and Lost Basin Mining Districts of Mohave County, Arizona that will be of benefit to miners and prospectors. Also included is a significant body of information on the gold mines and prospects of this portion of Arizona. This volume is lavishly illustrated with rare photos and mining maps. 8.5" X 11", 188 ppgs. Retail Price: $19.99

Mines of the Jerome and Bradshaw Mountains of Arizona - This important publication on Arizona Mining has not been available for ninety years. This volume contains rare insights into the geology and ore deposits of the Jerome and Bradshaw Mountains of Arizona that will be of benefit to miners and prospectors who work those areas. Included is a significant body of information on the mines and prospects of the Verde, Black Hills, Cherry Creek, Prescott, Walker, Groom Creek, Hassayampa, Bigbug, Turkey Creek, Agua Fria, Black Canyon, Peck, Tiger, Pine Grove, Bradshaw, Tintop, Humbug and Castle Creek Mining Districts. This volume is lavishly illustrated with rare photos and mining maps. 8.5" X 11", 218 ppgs. Retail Price: $19.99

The Ajo Mining District of Pima County Arizona - This important publication on Arizona Mining has not been available for nearly seventy years. This volume contains rare insights into the geology and mineralization of the Ajo Mining District in Pima County, Arizona and in particular the famous New Cornelia Mine. 8.5" X 11", 126 ppgs. Retail Price: $11.99

Mining in the Santa Rita and Patagonia Mountains of Arizona - Originally published in 1915, this important publication on Arizona Mining has not been available for nearly a century. Included are rare insights into hundreds of gold, silver, copper and other mines in this famous Arizona mining area. Details include the locations, geology, history, production and other facts of the mines of this region. 8.5" X 11", 394 ppgs. Retail Price: $24.99

Mining in the Bisbee Quadrangle of Arizona - Originally published in 1906, this important publication on Arizona Mining has not been available for nearly a century. Included are rare insights into hundreds of gold, silver, copper and other mines in this famous Arizona mining area. Details include the locations, geology, history, production and other facts of the mines of this important mining region. 8.5" X 11", 188 ppgs. Retail Price: $14.99

Placer Gold Mining in Arizona - Unavailable since 1922, this hard to find publication includes valuable insights into the placer gold mines of the Arizona. Originally released as "Placer Gold of Arizona", despite its small size, this publication includes important details into the history, geology and locations of the well known placer gold mines and prospects in the State of Arizona. 48 ppgs, 8.99

Gold and Copper Mining near Payson, Arizona - Written in 1915, this hard to find publication includes valuable insights into the gold and copper mining industry of Arizona. Highlighted here are the gold and copper mines near Payson, Arizona. 68 ppgs, 8.99

Lode Gold Mining in Arizona - Unavailable since 1934, this hard to find publication, originally released as "Arizona Lode Gold Mines and Gold Mining" includes valuable insights into the gold mining industry of Arizona. Included are valuable insights into over 150 hardrock gold mines in over 30 different mining districts in Arizona. 278 ppgs, 21.99

Mining in the Dragoon Quadrangle of Cochise County, Arizona - Unavailable since 1964, this hard to find publication includes valuable insights into the mines of the Dragoon Quadrangle Mining Region. The publication includes important details into the history, geology and locations of the well known mines and prospects in this famous mining region of Arizona. 224 ppgs., 19.99

Directory of Operating Mines in Arizona in 1915 - Unavailable since 1916, this hard to find publication includes valuable insights into the mines of Arizona. This small publication includes a complete list of the mines that were operating in the State of Arizona during 1915 and includes details such as general location, owners and some basic facts about each mining operation. 52 ppgs. 8.99

Arizona Ore Deposits - Unavailable since 1938, this hard to find publication includes valuable insights into some ore deposits of Arizona. Included are valuable insights into the formation and characteristics of valuable ore deposits in the Jerome, Miami, Inspiration, Clifton, Morenci, Ray, Ajo, Eureka, Tombstone and Magma mining districts. Included are details into some of the major gold, silver and copper mines of these important Arizona mining areas. 160 ppgs, 14.99

Montana Mining Books

A History of Butte Montana: The World's Greatest Mining Camp - First published in 1900 by H.C. Freeman, this important publication sheds a bright light on one of the most important mining areas in the history of The West. Together with his insights, as well as rare photographs of the periods, Harry Freeman describes Butte and its vicinity from its early beginnings, right up to its flush years when copper flowed from its mines like a river. At the time of publication, Butte, Montana was known worldwide as "The Richest Mining Spot On Earth" and produced not only vast amounts of copper, but also silver, gold and other metals from its mines. Freeman illustrates, with great detail, the most important mines in the vicinity of Butte, providing rare details on their owners, their history and most importantly, how the mines operated and how their treasures were extracted. Of particular interest are the dozens of rare photographs that depict mines such as the famous Anaconda, the Silver Bow, the Smoke House, Moose, Paulin, Buffalo, Little Minah, the Mountain Consolidated, West Greyrock, Cora, the Green Mountain, Diamond, Bell, Parnell, the Neversweat, Nipper, Original and many others. **8.5" X 11", 142 ppgs. Retail Price: $12.99**

The Butte Mining District of Montana - This important publication on Montana Mining has not been available for over a century. Included are rare insights into the gold, copper and silver mines of Butte, Montana together with hard to find maps and photographs. Some of the topics include the early history of gold, silver and copper mining in the Butte area, insight into the geology of its mining areas, the local distribution of gold, silver and copper ores, as well their composition and how to identify them. Also included are detailed facts about the mines in the Butte Mining District, including the famous Anaconda Mine, Gagnon, Parrot, Blue Vein, Moscow, Poulin, Stella, Buffalo, Green Mountain, Wake Up Jim, the Diamond-Bell Group, Mountain Consolidated, East Greyrock, West Greyrock, Snowball, Corra, Speculator, Adirondack, Miners Union, the Jessie-Edith May Group, Otisco, Iduna, Colorado, Lizzie, Cambers, Anderson, Hesperus, Preferencia and dozens of others. **8.5" X 11", 298 ppgs. Retail Price: $24.99**

Mines of the Helena Mining Region of Montana - This important publication on Montana Mining has not been available for over a century. Included are rare insights into the gold, copper and silver mines of the vicinity of Helena, Montana, including the Marysville Mining District, Elliston Mining District, Rimini Mining District, Helena Mining District, Clancy Mining District, Wickes Mining District, Boulder and Basin Mining Districts and the Elkhorn Mining District. Some of the topics include the early history of gold, silver and copper mining in the Helena area, insight into the geology of its mining areas, the local distribution of gold, silver and copper ores, as well their composition and how to identify them. Also included are detailed facts, history, geology and locations of over one hundred gold, silver and copper mines in the area . **8.5" X 11", 162 ppgs, Retail Price: $14.99**

Mines and Geology of the Garnet Range of Montana - This important publication on Montana Mining has not been available for over a century. Included are rare insights into the gold, copper and silver mines of the vicinity of this important mining area of Montana. Some of the topics include the early history of gold, silver and copper mining in the Garnet Mountains, insight into the geology of its mining areas, the local distribution of gold, silver and copper ores, as well their composition and how to identify them. Also included are detailed facts, history, geology and locations of numerous gold, silver and copper mines in the area . **8.5" X 11", 100 ppgs, Retail Price: $11.99**

Mines and Geology of the Philipsburg Quadrangle of Montana - This important publication on Montana Mining has not been available for over a century. Included are rare insights into the gold, copper and silver mines of the vicinity of this important mining area of Montana. Some of the topics include the early history of gold, silver and copper mining in the Philipsburg Quadrangle, insight into the geology of its mining areas, the local distribution of gold, silver and copper ores, as well their composition and how to identify them. Also included are detailed facts, history, geology and locations of over one hundred gold, silver and copper mines in the area **8.5" X 11", 290 ppgs, Retail Price: $24.99**

Geology of the Marysville Mining District of Montana - Included are rare insights into the mining geology of the Marysville Mining District. Some of the topics include the early history of gold, silver and copper mining in the area, insight into the geology of its mining areas, the local distribution of gold, silver and copper ores, as well their composition and how to identify them. Also included are detailed facts, history, geology and locations of gold, silver and copper mines in the area **8.5" X 11", 198 ppgs, Retail Price: $19.99**

The Geology and Mines of Northern Idaho and North Western Montana- See listing under Idaho.

The History of Gold Dredging in Montana - Unavailable since 1916, this important publication was originally published by the Us Bureau of Mines and has been unavailable for a century. A century and more ago, giant dredging machines dug in Montana's rivers and creeks in search of illusive golden riches. First appearing in California in the 1850's, gold dredges finally reached their peak of development in Siberia and New Zealand before becoming popular again in the United States. This book offers a unique historical perspective on the gold dredges that once operated in Montana. This book on Montana mining history is lavishly illustrated with dozens of rare historic photos gold dredges that once operated in Montana, as well as hard to locate plans on how these dredges were designed. 120 ppgs., 11.99

Nevada Mining Books

The Bull Frog Mining District of Nevada - Unavailable since 1910, this publication was originally compiled by the United States Department of Interior. This volume also includes important insights into the geologic formations, faults and other aspects of economic geology in this Nevada mining district. Of particular interest are the fine details on many mines in the area, including their locations, histories, development and mineralization. Some of the mines featured include the National Bank Mine, Providence, Gibraltor, Tramps, Denver, Original Bullfrog, Gold Bar, Mayflower, Homestake-King and other mines and prospects. **8.5" X 11", 152 ppgs, Retail Price: $14.99**

History of the Comstock Lode - Unavailable since 1876, this publication was originally released by John Wiley & Sons. This volume also includes important insights into the famous Comstock Lode of Nevada that represented the first major silver discovery in the United States. During its spectacular run, the Comstock produced over 192 million ounces of silver and 8.2 million ounces of gold. Not only did the Comstock result in one of the largest mining rushes in history and yield immense fortunes for its owners, but it made important contributions to the development of the State of Nevada, as well as neighboring California. Included here are important details on not only the early development and history of the Comstock, but also rare early insight into its mines, ore and its geology.**8.5" X 11", 244 ppgs, Retail Price: $19.99**

The Pioche Mining District of Nevada - First published in 1932, it has been unavailable for over a century and sheds important light on the mining history of Nevada. Some of the topics include the history of mining in this district, as well as the characteristics of its mineral and ore deposits. Also included are insights into the history, production, characteristics and locations of numerous mines in the area. Some of the mines include the Combined Metals, Pioche, Ely Valley, No. 10, Poorman, Wide Awake, Alps, Prince, Virginia Louise, Half Moon, Abe Lincoln, Fairview, Bristol Silver, National, Vesuvius, Inman, Tempest, Hillside, Jackrabbit, Lucky Star, Fortuna, Mendha, Manhattan, Hamburg, Comet, Lyndon and others. 108 ppgs 10.99

The Yerington Mining District of Nevada - First published in 1932, it has been unavailable for over a century and sheds important light on the mining history of Nevada. Some of the topics include the history of mining in this district, as well as the characteristics of its mineral and ore deposits. Also included are insights into the history, production, characteristics and locations of numerous mines in the area. Some of the mines include the Bluestone, Mason Valley, Malachite, McConnell, Greenwood, Western Nevada, Ludwig, Douglas Hill, Casting Copper, Montana-Yerington, Empire, Jim Beatty, Terry and McFarland, Blue Jay and others. 92 ppgs, 10.99

The Genesis of the Ores of Tonopah Nevada - Unavailable since 1918, this hard to find publication includes valuable insights into the gold mines around Tonopah, Nevada. The publication includes important details into the geology of mines in the Tonopah Mining District of Nevada. 90 ppgs, 10.99

Mining Camps of Elko, Lander and Eureka Counties Nevada - Unavailable since 1910, this hard to find publication includes valuable insights into the mining camps of Elko, Lander and Eureka Counties, Nevada. The publication includes important details into the history of mines and mining in these three Nevada counties. 154 ppgs, 12.99

Ore Deposits of the Bullfrog Quadrangle - Unavailable since 1964 and released as "Geology of Bullfrog Quadrangle and Ore Deposits Related to Bullfrog Hills Caldera, Nye County, Nevada and Inyo County, California". The publication includes important details into the geology of mines in the Bullfrog Quadrangle of Nye County, Nevada and Inyo County, California. 52 ppgs, 9.99

Mining in Eureka County Nevada - Unavailable since 1879, this hard to find publication includes valuable insights into the early mining history off Eureka County, Nevada. The publication includes important details into the early history of the mines of Eureka County, as well as their development, production and how their ores were treated. Also included are details on the 1872 Mining Act, as well as the local rules, regulations and customs of the miners in Eureka County.134 ppgs, 12.99

Colorado Mining Books

Ores of The Leadville Mining District - Unavailable since 1926, this publication was originally compiled by the United States Department of Interior. This volume also includes important insights into the ores and mineralization of the Leadville Mining District in Colorado. Topics include historic ore prospecting methods, local geology, insights into ore veins and stockworks, the local trend and distribution of ore channels, reverse faults, shattered rock above replacement ore bodies, mineral enrichment in oxidized and sulphide zones and more. **8.5″ X 11″, 66 ppgs, Retail Price: $8.99**

Mining in Colorado - Unavailable since 1926, this publication was originally compiled by the United States Department of Interior. This volume also includes important insights into the mining history of Colorado from its early beginnings in the 1850's right up to the mid 1920's. Not only is Colorado's gold mining heritage included, but also its silver, copper, lead and zinc mining industry. Each mining area is treated separately, detailing the development of Colorado's mines on a county by county basis. **8.5″ X 11″, 284 ppgs, Retail Price: $19.99**

Gold Mining in Gilpin County Colorado - Unavailable since 1876, this publication was originally compiled by the Register Steam Printing House of Central City, Colorado. A rare glimpse at the gold mining history and early mines of Gilpin County, Colorado from their first discovery in the 1850's up to the "flush years" of the mid 1870's. Of particular interest is the history of the discovery of gold in Gilpin County and details about the men who made those first strikes. Special focus is given to the early gold mines and first mining districts of the area, many of which are not detailed in other books on Colorado's gold mining history. **8.5″ X 11″, 156 ppgs, Retail Price: $12.99**

Mining in the Gold Brick Mining District of Colorado - Important insights into the history of the Gold Brick Mining District, as well as its local geography and economic geology. Also included are the histories and locations of historic mines in this important Colorado Mining District, including the Cortland, Carter, Raymond, Gold Links, Sacramento, Bassick, Sandy Hook, Chronicle, Grand Prize, Chloride, Granite Mountain, Lucille, Gray Mountain, Hilltop, Maggie Mitchell, Silver Islet, Revenue, Roosevelt, Carbonate King and others. In addition to hardrock mining, are also included are details on gold placer mining in this portion of Colorado. **8.5″ X 11″, 140 ppgs, Retail Price: $12.99**

Ore Deposits of the London Fault of Colorado - First published in 1941, it has been unavailable since those days and sheds important light on the mines and mineral deposits of the London Fault in Central Colorado's Alma Mining District. This publication sheds important light on the gold veins and lead-silver deposits of the Alma Mining District. Included are geologic details on the London Mine, American Mine, Havigorst Tunnel, Ophir Mine, Mosher Tunnel, London-Butte Mine, Venture Shaft, Hard-To-Beat Mine, Oliver Twist Tunnel, Sacramento Mine, Mudsill Mine, Sherwood Mine, Wagner, Barcoe Tunnel and other mines in this important mining region. 110 ppgs., 10.99

The Mines of Colorado - First published in 1867, it has been unavailable since those days and sheds important light on Colorado's early mining history. Written shortly after the events took place, this publication sheds important light on the Pike's Peak Gold Rush, the discovery of gold on Ralston Creek and Dry Creek in the 1850's, as well as details on the first wave of miners into Colorado and their trials and tribulations as they crossed the Great Plains. Also included are details on early discoveries of lode gold in the mountainous regions of Colorado, details on the early mines hardrock and placer mines, and much more. It is a veritable treasure trove on Colorado's early mining history and will be of great importance to anyone who is interested in the mining of gold or other minerals in Colorado, as well as those interested in the history of the state. 478 ppgs., 29.99

The La Plata Mining District of Colorado - Originally titled "Geology and Ore Deposits in the Vicinity of the La Plata District of Colorado" and first published in 1949, it has been unavailable since those days and sheds important light on the mines and mineral deposits of the La Plata Mining District of Colorado. 214 ppgs., 19.99

Washington Mining Books

The Republic Mining District of Washington - Unavailable since 1910, this important publication was originally published by the Washington Geologic Survey and has been unavailable for a century. Topics include the geology, rock formations and the formation of ore deposits in this important mining area of Washington State. Also included are hard to find details on the geology, history and locations of dozens of mines in the area. Some of the mines featured include the New Republic Mine, Ben Hur, Morning Glory, the South Republic Mine, Quilp, Surprise, Black Tail, Lone Pine, San Poil, Mountain Lion, Tom Thumb, Elcaliph and many others. **8.5" X 11", 94 ppgs, Retail Price: $10.99**

The Myers Creek and Nighthawk Mining Districts of Washington - Unavailable since 1911, this important publication was originally published by the Washington Geologic Survey and has been unavailable for a century. Topics include the geology, rock formations and the formation of ore deposits in these important mining areas of Washington State. Also included are hard to find details on the geology, history and locations of dozens of mines in the area. Some of the mines featured include the Grant Mine, Monterey, Nip and Tuck, Myers Creek, Number Nine, Neutral, Rainbow, Aztec, Crystal Butte, Apex, Butcher Boy, Molson, Mad River, Olentangy, Delate, Kelsey, Golden Chariot, Okanogan, Ohio, Forty-Ninth Parallel, Nighthawk, Favorite, Little Chopaka, Summit, Number One, California, Peerless, Caaba, Prize Group, Ruby, Mountain Sheep, Golden Zone, Rich Bar, Similkameen, Kimberly, Triune, Hiawatha, Trinity, Hornsilver, Maquae, Bellevue, Bullfrog, Palmer Lake, Ivanhoe, Copper World and many others. **8.5" X 11", 136 ppgs, Retail Price: $12.99**

The Blewett Mining District of Washington - Unavailable since 1911, this important publication was originally published by the Washington Geologic Survey and has been unavailable for a century. Topics include the geology, rock formations and the formation of ore deposits in this important mining area of Washington State. Also included are hard to find details on the geology, history and locations of dozens of mines in the area. Some of the mines featured include the Washington Meteor, Alta Vista, Pole Pick, Blinn, North Star, Golden Eagle, Tip Top, Wilder, Golden Guinea, Lucky Queen, Blue Bell, Prospect, Homestake, Lone Rock, Johnson, and others. **8.5" X 11", 134 ppgs, Retail Price: $12.99**

Silver Mining In Washington - Unavailable since 1955, this important publication was originally published by the Washington Geologic Survey. Featured are the hard to find locations and details pertaining to Washington's silver mines. **8.5" X 11", 180 ppgs, Retail Price: $15.99**

The Mines of Snohomish County Washington - Unavailable since 1942, this important publication was originally published by the Washington Geologic Survey and has been unavailable for seventy years. Featured are details on a large number of gold, silver, copper, lead and other metallic mineral mines. Included are the locations of each historic mine, along with information on the commodity produced. **8.5" X 11", 98 ppgs, Retail Price: $10.99**

The Mines of Chelan County Washington - Unavailable since 1943, this important publication was originally published by the Washington Geologic Survey and has been unavailable for seventy years. Featured are details on a large number of gold, silver, copper, lead and other metallic mineral mines. Included are the locations of each historic mine, along with information on the commodity. **8.5" X 11", 88 ppgs, Retail Price: $9.99**

Metal Mines of Washington - Unavailable since 1921, this important publication was originally published by the Washington Geologic Survey and has been unavailable for nearly ninety years. Widely considered a masterpiece on the Washington Mining Industry, "Metal Mines of Washington" sheds light on the important details of Washington's early mining years. Featured are details on hundreds of gold, silver, copper, lead and other metallic mineral mines. Included are hard to find details on the mineral resources of this state, as well as the locations of historic mines. Lavishly illustrated with maps and historic photos and complete with a glossary to explain any technical terms found in the text, this is one of the most important works on mining in the State of Washington. No prospector or miner should be without it if they are interested in mining in Washington. **8.5" X 11", 396 ppgs, Retail Price: $24.99**

Gem Stones In Washington - Unavailable since 1949, this important publication was originally published by the Washington Geologic Survey and has been unavailable since first published. Included are details on where to find naturally occurring gem stones in the State of Washington, including quartz crystal, amethyst, smoky quartz, milky quartz, agates, bloodstone, carnelian, chert, flint, jasper, onyx, petrified wood, opal, fire opal, hyalite and others. **8.5" X 11", 54 ppgs, Retail Price: $8.99**

The Covada Mining District of Washington - Unavailable since 1913, this important publication was originally published by the Washington Geologic Survey and has been unavailable for a century. Topics include the geology, rock formations and the formation of ore deposits in this important mining area of Washington State. Also included are hard to find details on the geology, history and locations of dozens of mines in the area. Some of the mines featured include the Admiral, Advance, Algonkian, Big Bug, Big Chief, Big Joker, Black Hawk, Black Tail, Black Thorn, Captain, Cherokee Strip, Colorado, Dan Patch, Dead Shot, Etta, Good Ore, Greasy Run, Great Scott, Idora, IXL, Jay Bird, Kentucky Bell, King Solomon, Laurel, Laura S, Little Jay, Meteor, Neglected, Northern Light, Old Nell, Plymouth Rock, Polaris, Quandary, Reserve, Shoo Fly, Silver Plume, Three Pines, Vernie, White Rose and dozens of others. **8.5" X 11", 114 ppgs, Retail Price: $10.99**

The Index Mining District of Washington - Unavailable since 1912, this important publication was originally published by the Washington Geologic Survey and has been unavailable for a century. Topics include the geology, rock formations and the formation of ore deposits in this important mining area of Washington State. Also included are hard to find details on the geology, history and locations of dozens of mines in the area. Some of the mines featured include the Sunset, Non-Pareil, Ethel Consolidated, Kittaning, Merchant, Homestead, Co-operative, Lost Creek, Uncle Sam, Calumet, Florence-Rae, Bitter Creek, Index Peacock, Gunn Peak, Helena, North Star, Buckeye. Copper Bell, Red Cross and others. **8.5" X 11", 114 ppgs, Retail Price: $11.99**

Mining & Mineral Resources of Stevens County Washington - Unavailable since 1920, this important publication was originally published by the Washington Geologic Survey and has been unavailable for a century. Topics include the geology, rock formations and the formation of ore deposits in these important mining areas of Washington State. Also included are hard to find details on the geology, history and locations of hundreds of mines in the area. **8.5" X 11", 372 ppgs, Retail Price: $24.99**

The Mines and Geology of the Loomis Quadrangle Okanogan County, Washington - Unavailable since 1972, this important publication was originally published by the Washington Geologic Survey and has been unavailable for a century. Topics include the geology, rock formations and the formation of ore deposits in this important mining area of Washington State. Also included are hard to find details on the geology, history and locations of dozens of gold, copper, silver and other mines in the area. **8.5" X 11", 150 ppgs, Retail Price: $12.99**

The Conconully Mining District of Okanogan County Washington - Unavailable since 1973, this important publication was originally published by the Washington Geologic Survey and has been unavailable for a century. Topics include the geology, rock formations and the formation of ore deposits in this important mining area of Washington State, which also includes Salmon Creek, Blue Lake and Galena. Also included are hard to find details on the geology, mining history and locations of dozens of mines in the area. Some of the mines include Arlington, Fourth of July, Sonny Boy, First Thought, Last Chance, War Eagle-Peacock, Wheeler, Mohawk, Lone Star, Woo Loo Moo Loo, Keystone, Hughes, Plant-Callahan, Johnny Boy, Leuena, Gubser, John Arthur, Tough Nut, Homestake, Key and many others **8.5" X 11", 68 ppgs, Retail Price: $8.99**

Wyoming Mining Books

Mining in the Laramie Basin of Wyoming - Unavailable since 1909, this publication was originally compiled by the United States Department of Interior. Also included are insights into the mineralization and other characteristics of this important mining region, especially in regards to coal, limestone, gypsum, bentonite clay, cement, sand, clay and copper. **8.5" X 11", 104 ppgs, Retail Price: $11.99**

New Mexico Mining Books

The Mogollon Mining District of New Mexico - Unavailable since 1927, this important publication was originally published by the US Department of Interior and has been unavailable for 80 years. Topics include the geology, rock formations and the formation of ore deposits in this important mining area in New Mexico. Of particular focus is information on the history and production of the ore deposits in this area, their form and structure, vein filling, their paragenesis, origins and ore shoots, as well as oxidation and supergene enrichment. Also included are hard to find details, including the descriptions and locations of numerous gold, silver and other types of mines, including the Eureka, Pacific, South Alpine, Great Western, Enterprise, Buffalo, Mountain View, Floride, Gold Dust, Last Chance, Deadwood, Confidence, Maud S., Deep Down, Little Fanney, Trilby, Johnson, Alberta, Comet, Golden Eagle, Cooney, Queen, the Iron Crown, Eberle, Clifton, Andrew Jackson mine, Mascot and others. **8.5" X 11", 144 ppgs, Retail Price: $12.99**

The Percha Mining District of Kingston New Mexico - Unavailable since 1883, this important publication was originally published by the Kingston Tribune and has been unavailable for over one hundred and thirty five years. Having been written during the earliest years of gold and silver mining in the Percha Mining District, unlike other books on the subject, this work offers the unique perspective of having actually been written while the early mining history of this area was still being made. In fact, the work was written so early in the development of this area that many of the notable mines in the Percha District were less than a few years old and were still being operated by their original discoverers with the same enthusiasm as when they were first located. Included are hard to find details on the very earliest gold and silver mines of this important mining district near Kingston in Sierra County, New Mexico. **8.5" X 11", 68 ppgs, Retail Price: $9.99**

East Coast Mining Books

<u>The Gold Fields of the Southern Appalachians</u> - Unavailable since 1895, this important publication was originally published by the US Department of Interior and has been unavailable for nearly 120 years. Topics include the geology, rock formations and the formation of ore deposits in this important mining area of the American South. Of particular focus is information on the history and statistics of the ore deposits in this area, their form and structure and veins. Also included are details on the placer gold deposits of the region. The gold fields of the Georgian Belt, Carolinian Belt and the South Mountain Mining District of North Carolina are all treated in descriptive detail. Included are hard to find details, including the descriptions and locations of numerous gold mines in Georgia, North Carolina and elsewhere in the American South. Also included are details on the gold belts of the British Maritime Provinces and the Green Mountains. **8.5" X 11", 104 ppgs, Retail Price: $9.99**

Gold Rush Tales Series

<u>Millions in Siskiyou County Gold</u> - In this first volume of the "Gold Rush Tales" series, leading mining historian and editor Kerby Jackson, introduces us to the story of how millions of dollars worth of gold was discovered in Siskiyou County during the California Gold Rush. Lavishly illustrated with photos from the 19th Century, this hard to find information was first published in 1897 and sheds important light onto the gold rush era in Siskiyou County, California and the experiences of the men who dug for the gold and actually found it. **8.5" X 11", 82 ppgs, Retail Price: $9.99**

<u>The California Rand in the Days of '49</u> - In this second volume of the "Gold Rush Tales" series, leading mining historian and editor Kerby Jackson, introduces us to four tales from the California Gold Rush. Lavishly illustrated with photos from the 19th Century, this hard to find information was first published in 1890's and includes the stories of "California's Rand", details about Chinese miners, how one early miner named Baker struck it rich and also the story of Alphonzo Bowers, who invented the first hydraulic gold dredge. **8.5" X 11", 54 ppgs, Retail Price: $9.99**

More Mining Books

<u>Prospecting and Developing A Small Mine</u> - Topics covered include the classification of varying ores, how to take a proper ore sample, the proper reduction of ore samples, alluvial sampling, how to understand geology as it is applied to prospecting and mining, prospecting procedures, methods of ore treatment, the application of drilling and blasting in a small mine and other topics that the small scale miner will find of benefit. **8.5" X 11", 112 ppgs, Retail Price: $11.99**

<u>Timbering For Small Underground Mines</u> - Topics covered include the selection of caps and posts, the treatment of mine timbers, how to install mine timbers, repairing damaged timbers, use of drift supports, headboards, squeeze sets, ore chute construction, mine cribbing, square set timbering methods, the use of steel and concrete sets and other topics that the small underground miner will find of benefit. This volume also includes twenty eight illustrations depicting the proper construction of mine timbering and support systems that greatly enhance the practical usability of the information contained in this small book. **8.5" X 11", 88 ppgs. Retail Price: $10.99**

<u>Timbering and Mining</u> - A classic mining publication on Hard Rock Mining by W.H. Storms. Unavailable since 1909, this rare publication provides an in depth look at American methods of underground mine timbering and mining methods. Topics include the selection and preservation of mine timbers, drifting and drift sets, driving in running ground, structural steel in mine workings, timbering drifts in gravel mines, timbering methods for driving shafts, positioning drill holes in shafts, timbering stations at shafts, drainage, mining large ore bodies by means of open cuts or by the "Glory Hole" system, stoping out ore in flat or low lying veins, use of the "Caving System", stoping in swelling ground, how to stope out large ore bodies, Square Set timbering on the Comstock and its modifications by California miners, the construction of ore chutes, stoping ore bodies by use of the "Block System", how to work dangerous ground, information on the "Delprat System" of stoping without mine timbers, construction and use of headframes and much more. This volume provides a reference into not only practical methods of mining and timbering that may be employed in narrow vein mining by small miners today, but also rare insights into how mines were being worked at the turn of the 19th Century. **8.5" X 11", 288 ppgs. Retail Price: $24.99**

A Study of Ore Deposits For The Practical Miner - Mining historian Kerby Jackson introduces us to a classic mining publication on ore deposits by J.P. Wallace. First published in 1908, it has been unavailable for over a century. Included are important insights into the properties of minerals and their identification, on the occurrence and origin of gold, on gold alloys, insights into gold bearing sulfides such as pyrites and arsenopyrites, on gold bearing vanadium, gold and silver tellurides, lead and mercury tellurides, on silver ores, platinum and iridium, mercury ores, copper ores, lead ores, zinc ores, iron ores, chromium ores, manganese ores, nickel ores, tin ores, tungsten ores and others. Also included are facts regarding rock forming minerals, their composition and occurrences, on igneous, sedimentary, metamorphic and intrusive rocks, as well as how they are geologically disturbed by dikes, flows and faults, as well as the effects of these geologic actions and why they are important to the miner. Written specifically with the common miner and prospector in mind, the book will help to unlock the earth's hidden wealth for you and is written in a simple and concise language that anyone can understand. **8.5″ X 11″, 366 ppgs. Retail Price: $24.99**

Mine Drainage - Unavailable since 1896, this rare publication provides an in depth look at American methods of underground mine drainage and mining pump systems. This volume provides a reference into not only practical methods of mining drainage that may be employed in narrow vein mining by small miners today, but also rare insights into how mines were being worked at the turn of the 19th Century. **8.5″ X 11″, 218 ppgs. Retail Price: $24.99**

Fire Assaying Gold, Silver and Lead Ores - Unavailable since 1907, this important publication was originally published by the Mining and Scientific Press and was designed to introduce miners and prospectors of gold, silver and lead to the art of fire assaying. Topics include the fire assaying of ores and products containing gold, silver and lead; the sampling and preparation of ore for an assay; care of the assay office, assay furnaces; crucibles and scorifiers; assay balances; metallic ores; scorification assays; cupelling; parting' crucible assays, the roasting of ores and more. This classic provides a time honored method of assaying put forward in a clear, concise and easy to understand language that will make it a benefit to even beginners. **8.5″ X 11″, 96 ppgs. Retail Price: $11.99**

Methods of Mine Timbering - Originally published in 1896, this important publication on mining engineering has not been available for nearly a century. Included are rare insights into historical methods of timbering structural support that were used in underground metal mines during the California that still have a practical application for the small scale hardrock miner of today. **8.5″ X 11″, 94 ppgs. Retail Price: $10.99**

The Enrichment of Copper Sulfide Ores - First published in 1913, it has been unavailable for over a century. Topics include the definition and types of ore enrichment, the oxidation of copper ores, the precipitation of metallic sulfides. Also included are the results of dozens of lab experiments pertaining to the enrichment of sulfide ores that will be of interest to the practical hard rock mine operator in his efforts to release the metallic bounty from his mine's ore. **8.5″ X 11″, 92 ppgs. Retail Price: $9.99**

A Study of Magmatic Sulfide Ores - Unavailable since 1914, this rare publication provides an in depth look at magmatic sulfide ores. Some of the topics included are the definition and classification of magmatic ores, descriptions of some magmatic sulfide ore deposits known at the time of publication including copper and nickel bearing pyrrohitic ore bodies, chalcopyrite-bornite deposits, pyritic deposits, magnetite-ileminite deposits, chromite deposits and magmatic iron ore deposits. Also included are details on how to recognize these types of ore deposits while prospecting for valuable hardrock minerals. **8.5″ X 11″, 138 ppgs. Retail Price: $11.99**

The Cyanide Process of Gold Recovery - Unavailable since 1894 and released under the name "The Cyanide Process: Its Practical Application and Economical Results", this rare publication provides an in depth look at the early use of cyanide leaching for gold recovery from hardrock mine ores. This volume provides a reference into the early development and use of cyanide leaching to recover gold. **8.5″ X 11″, 162 ppgs. Retail Price· $14.99**

California Gold Milling Practices - Unavailable since 1895 and released under the name "California Gold Practices", this rare publication provides an in depth look at early methods of milling used to reduce gold ores in California during the late 19th century. This volume provides a reference into the early development and use of milling equipment during the earliest years of the California Gold Rush up to the age of the Industrial Revolution. Much of the information still applies today and will be of use to small scale miners engaging in hardrock mining. **8.5″ X 11″, 104 ppgs. Retail Price: $10.99**

Leaching Gold and Silver Ores With The Plattner and Kiss Processes - Mining historian Kerby Jackson introduces us to a classic mining publication on the evaluation and examination of mines and prospects by C.H. Aaron. First published in 1881, it has been unavailable for over a century and sheds important light on the leaching of gold and silver ores with the Plattner and Kiss processes. **8.5″ X 11″, 204 ppgs. Retail Price: $15.99**

The Metallurgy of Lead and the Desilverization of Base Bullion - First published in 1896, it has been unavailable for over a century and sheds important light on the the recovery of silver from lead based ores. Some of the topics include the properties of lead and some of its compounds, lead ores such as galenite, anglesite, cerussite and others, the distribution of lead ores throughout the United States and the sampling and assaying of lead ores. Also covered is the metallurgical treatment of lead ores, as well as the desilverization of lead by the Pattinson Process and the Parkes Process. Hofman's text has long been considered one of the most important early works on the recovery of silver from lead based ores. 8.5" X 11", 452 ppgs. **Retail Price: $29.99**

Ore Sampling For Small Scale Miners - First published in 1916, it has been unavailable for over a century and sheds important light on historic methods of ore sampling in hardrock mines. Topics include how to take correct ore samples and the conditions that affect sampling, such as their subdivision and uniformity. Particular detail is given to methods of hand sampling ore bodies by grab sample, pipe sample and coning, as well as sampling by mechanical methods. Also given are insights into the screening, drying and grinding processes to achieve the most consistent sample results and much more. 8.5" X 11", 124 ppgs. **Retail Price: $12.99**

The Extraction of Silver, Copper and Tin from Ores - First published in 1896, it has been unavailable for over a century and sheds important light on how historic miners recovered silver, copper and tin from their mining operations. The book is split into three sections, including a discussion on the Lixiviation of Silver Ores, the mining and treatment of copper ores as practiced at Tharsis, Spain and the smelting of tin as it was practiced by metallurgists at Pulo Brani, Singapore. Also included is an overview and analysis of these historic metal recovery methods that will be of benefit to those interested in the extraction of silver, copper and tin from small mines. 8.5" X 11", 118 ppgs. **Retail Price: $14.99**

The Roasting of Gold and Silver Ores - First published in 1880, it has been unavailable for over a century and sheds important light on how historic miners recovered gold and silver rom their mining operations. Topics include details on the most important silver and free milling gold ores, methods of desulphurization of ores, methods of deoxidation, the chlorination of ores, methods and details on roasting gold and silver ores, notes on furnaces and more. Also included are details on numerous methods of gold and silver recovery, including the Ottokar Hofman's Process, the Patera Process, Kiss Process, Augustin Process, Ziervogel Process and others. 8.5" X 11", 178 ppgs. **Retail Price: $19.99**

The Examination of Mines and Prospects - First published in 1912, it has been unavailable for over a century and sheds important light on how to examine and evaluate hardrock mines, prospects and lode mining claims. Sections include Mining Examinations, Structural Geology, Structural Features of Ore Deposits, Primary Ores and their Distribution, Types of Primary Ore Deposits, Primary Ore Shoots, The Primary Alteration of Wall Rocks, Alterations by Surface Agencies, Residual Ores and their Distribution, Secondary Ores and Ore Shoots and Vein Outcrops. This hard to find information is a must for those who are interested in owning a mine or who already own a lode mining claim and wish to succeed at quartz mining. 8.5" X 11", 250 ppgs. **Retail Price: $19.99**

Garnets: Their Mining, Milling and Utilization - First published in 1925, it has been unavailable since those days and sheds important light on the mining, milling and utilization of garnets. Included are details on the characteristics of garnets, where they are found and how they were mined. 78 ppgs, 10.99

Gemstones and Precious Stones of North America - Leading mining historian Kerby Jackson introduces us to a classic mining publication on the gems and precious stones of the United States, Canada and mexico. First published in 1890, it has been unavailable since those days and sheds important light on the gems and precious stones that may be found in North America. Included are chapters on diamonds, corundum, sapphire, ruby, topaz, emerald, disapore, spinel, turquoise, tourmaline, garnets, beyrl, peridot, zircon, quartz crystals, feldspars, pearls and many others. Included are details on where these gems and precious stones may be found throughout North America, as well as their characteristics. 360 ppgs, 24.99

Mining Camps and Mining Districts - First released in 1885 by Charles Howard Shinn under the title "Mining Camps: A Study in American Frontier Government", this publication offers a unique look at how early gold miners established their own forms of representative government during the California Gold Rush. Drawing on the the early mining codes of mideviel German miners in the Harz Mountains, on the mining customs of the Cornish tin miners and early Spanish mining laws introduced into California, the miners established the first governments in the American West. 340 ppgs, 24.99

BLM Field Handbook for Mineral Examiners - Leading mining historian Kerby Jackson introduces us to a classic mining publication on mine evaluation. First published in 1962, this work sheds important light on the techniques of BLM Mineral Examiners to perform validity on mining claims. 132 ppgs, 10.99

Six Months In The Gold Mines During The California Gold Rush - Unavailable since 1850, this important work is a first hand account of one "49'ers" personal experience during the great California Gold Rush, shedding important light on one of the most exciting periods in the history of not only California, but also the world. Compiled from journals written between 1847 and 1849 by E. Gould Buffum, a native of New York, "Six Months In The Gold Mines During The California Gold Rush" offers a rare look into the day to day lives of the people who came to California to work in her gold mines when the state was still a great frontier. **8.5" X 11", 290 ppgs. Retail Price: $19.99**

The Discovery of Gold in Australia - **First published in 1852, it has been unavailable since those days and sheds important light on Australia's gold mining history. Included are rare communications between British agents and the British Crown when gold was first discovered in Australia in 1851. This rare text contains hard to find details on Australia's first mining camps and Britain's early attempts to provide for the orderly regulation of gold mines in that part of the world. Also of interest are hard to find extracts of articles that appeared in the early colonial newspapers that did their best to report on Australia's gold rush as it took place.**
102 ppgs, 10.99